RADIOGRAPHY EXAM

**OTHER TITLES OF INTEREST FROM
LEARNINGEXPRESS**

Medical Assistant: Preparation for the CMA and RMA Exams
Becoming a Healthcare Professional
Health Occupations Entrance Exams

RADIOGRAPHY EXAM

LEARNINGEXPRESS®

NEW YORK

Library of Congress Cataloging-in-Publication
Radiography exam. — 1st ed.
        p. ; cm.
ISBN-13: 978-1-57685-731-1 (pbk. : alk. paper)
ISBN-10: 1-57685-731-X (pbk. : alk. paper)
1. Radiologic technologists—Examinations, questions, etc.   2. Radiologic technologists—Examinations—
Study guides. I. LearningExpress (Organization)
  [DNLM: 1. Radiography—Examination Questions.   2. Technology, Radiologic—Examination Questions.
WN 18.2]
  RM849.3.R33 2011
  616.07'57076—dc22
                                                2010040778

Printed in the United States of America

9 8 7 6 5 4 3 2 1

Developed by Print Matters, Inc., www.printmattersinc.com.

ISBN: 978-1-57685-731-1

For information on LearningExpress, other LearningExpress products,
or bulk sales, please write to us at:
  2 Rector Street
  26th Floor
  New York, NY 10006

Or visit us at:
  www.learnatest.com

CONTENTS ▶

CONTRIBUTORS ▶

Anthony Mungo, BS, RT (R), is a licensed radiologic technologist and currently serves as the Administrative and Technical Director of Radiology at New York Methodist Hospital in Brooklyn, New York. Mr. Mungo is responsible for providing leadership for 160 employees and services for a 590-bed academic institution. Prior to this role, Anthony served as a Regional Director of Imaging for several outpatient centers in the New York City Area. Over the past 10 years, Anthony has taught several courses in radiology discipline, including patient care, medical terminology, radiographic procedures, magnetic resonance imaging, and computed tomography cross-sectional anatomy. Mr. Mungo holds a bachelor's degree in Health Sciences and a degree in Radiologic Technology. Anthony is an active Advisory Board Member of the Advisory Board in Washington, D.C., and is a member of the American College of Forensic Examiners, Radiology Business Management Association, American College of Healthcare Executives, American Society of Radiologic Technologists, American Healthcare Radiology Administrators, and American Registry of Radiologic Technologists. Mungo has lectured at several large forums across the country.

Michael Ficorelli, RT (R), serves as Manager of Radiology and Advanced Imaging Services at the New York Methodist Hospital in Brooklyn, New York. Ficorelli is a well-versed instructor with over 10 years' experience teaching various courses throughout the radiography curriculum, including the Principles and Physics of Exposure, Radiographic Procedures, Positioning and Film Critique, Patient Care, and Radiation Protection. He also has direct clinical instruction with students. Ficorelli has 12 years of professional experience in the field as both a diagnostic radiographer and computed tomography technologist at both the New York Methodist Hospital and the New York Presbyterian Hospital/Weill Cornell Medical Center.

Kenneth Martinucci, MS, RT (R), is the Chairman of the Radiologic Technology Programs at Downstate Medical Center in Brooklyn, New York, and is the President of the New York Society of Radiologic Sciences. Professor Martinucci holds both a bachelor's and master's of science. With approximately 20 years of formal teaching experience, Professor Martinucci has taught a variety of courses in the discipline, including Radiographic Imaging, Patient Care, Health Care Ethics, Radiographic Procedures and Positioning, Darkroom Chemistry and Processor Quality Control, Pediatric Radiography, Radiographic Pathology, and Patient Care.

INTRODUCTION ▶

If you have made the decision to pursue a career as a radiologic technologist, you know that one of the requirements is to obtain certification and licensure by successfully passing the American Registry of Radiologic Technologists® (ARRT) examination. This test measures candidates' skills and knowledge to ensure that they can effectively perform the complicated and demanding responsibilities of being a radiologic technologist, either at a hospital or private imaging facility.

This book is designed to provide the practice, guidance, and support you will need to tackle the ARRT certification exam. It is not a substitute for your textbook. Rather, it is a supplement to the text and your classroom lessons. It contains the information you will need to improve your score on the ARRT exam in the shortest amount of time. This introduction serves as a guide for using this book effectively, so you get the most out of your time and effort.

Chapter 1 provides nuts and bolts information about the career of a radiologic technologist. Chapter 2 discusses the certification exam for radiologic technologists. Chapter 3 explains the secrets of test taking and how to improve your scores on multiple-choice exams through strategic planning.

LearningExpress has developed a system to help you prepare for your exam with proven test-taking strategies, including the following:

- becoming familiar with the format of the exam
- planning sufficient practice time prior to taking the exam
- learning how to overcome test anxiety
- learning critical test-taking skills, such as pacing yourself through the exam, using the process of elimination, and knowing when to guess
- getting enough rest and exercise before the test

The core of the book is found in Chapters 5 through 9, which cover the essential topics: radiation protection (Chapter 5), equipment operation and quality control (Chapter 6), image production and evaluation (Chapter 7), radiographic procedures and positioning (Chapter 8), and patient care and education (Chapter 9).

Chapters 4 and 10 are the heart of the book, containing the two full practice exams. The pretest in Chapter 4 is the place to start. Prepare yourself as if for the actual exam. Wear an accurate wristwatch, or use a stopwatch to allow yourself time to answer questions and skip those that hold you up. Before you read Chapters 5–9, take the first 200-question exam, allowing yourself 3.5 hours to complete it. Generally, the ARRT exam will have 220 questions. You will be graded on 200 questions; the other 20 are pilot questions randomly placed throughout the examination. The test taker will not know which questions are pilot questions. The examination is given at computer testing centers, and you can find one at a location near you. Each applicant must provide current photo identification prior to being admitted to the exam. Of course, cheating is strictly prohibited on the actual exam, and there is really no point in cheating on the practice tests in this book, since you will only be cheating yourself out of a chance to improve your knowledge.

Once you have completed the first practice examination, turn to the answers and explanations that follow. Score your exam carefully. Now, take note of the sections in which you did poorly. You will want to review these with extra attention. We suggest that your exam preparation include all of the core chapters (5–9), but you may want to concentrate on those sections in which you are weakest. Review each topic chapter here and in your textbook as well. At the end of each topic chapter, you will find a set of 100 review questions. Remember to time yourself as you answer them. Score these questions as you did the first practice exam to see how solid your knowledge of each topic is.

After completing the five review tests in Chapters 5–9, take the second practice exam in Chapter 10. Remember to time yourself. Score the exam, and identify which topics still need improvement.

Take the exams repeatedly. Remember, practice makes perfect.

CHAPTER 1 ▶ OVERVIEW OF A CAREER AS A RADIOGRAPHER

CHAPTER SUMMARY

This chapter will give you an understanding about the career of radiography. We will discuss what will be expected of you as an employee and what an employer will look for when selecting the right candidate for an open position. We will also discuss job descriptions, continuing education, resources, and some of the professional societies.

As you begin your career as a radiographer, you will come into contact with many new people in your field. It is always important to remember as a new employee of either a hospital or a private imaging facility that the expectations of your employer will be the same for all radiographers on the team. What does this mean? When interviewed by human resources, you will be asked for your qualifications, and they will provide you with a detailed list of responsibilities, also known as a job description, to which all employees with the same title must adhere. If you feel that the skill set that you possess and the education (didactic and clinical) you have received while in school qualify you for that position, then you will be considered a candidate. Upon hire, your new boss will expect you to have an understanding of positioning and anatomy, radiation exposure and protection, patient care, and many other subject matters that you (may) have learned in school. Your employer is only responsible for training you on the equipment, about exposure factors that pertain to the equipment, PACS (picture archiving and communication system), the radiology information system (RIS), and the hospital information system (HIS). While the overabundance of new information and technology may be overwhelming in the beginning, you are required to learn rather quickly. Onsite training is usually one to two weeks long, and in most circumstances a three-month employee performance evaluation will follow.

What is an employee performance evaluation? An evaluation is a tool employers use to review the performance of an employee. A three-month employee performance evaluation is common practice in health care and is most commonly referred to as a probation evaluation. This assessment will determine the employee's strengths, weaknesses, performance, and attendance over the past three months. The evaluation is based on your job description, essential duties, and responsibilities, which will be discussed shortly. If your performance is ever considered subpar or borderline acceptable, you may be placed on extended probation with a performance improvement plan.

How will you be graded? Your employer will grade your performance as being either satisfactory or unsatisfactory based on your job description, with a focus on three major categories, such as patient satisfaction, communication skills, and technical competency.

What can you do to get a good score? In any circumstance, it is always good practice as a new employee to ask questions. As you might have heard before, there is no such thing as a stupid question. If you ask your coworkers for suggestions or ways to figure out a particular problem that you may be facing, that will show that you care and you are enthusiastic to learn. Always greet your patients and coworkers with a smile and say good morning, afternoon, or evening to them. Ask them how they are doing and if there is anything you can do to make their experience a more pleasant one. Ask for assistance when needed. Be prompt when arriving to work and conscious of the time you leave. Ask your coworkers if they need help with difficult cases. All of these recommendations will show your employer that you care for your patients and coworkers, and this should reflect positively on your evaluation.

Will anyone else see my evaluation? An evaluation is personal information between management and the employee and should not be shared with coworkers. If, under any circumstance, you are not satisfied with the contents or results of your evaluation, you have the right to comment on the evaluator's report.

Job Description

The following is a list of job responsibilities for a radiographer.

Primary Responsibilities

- Review clinical information on each requisition for every patient, and prepare and execute the requested examination according to the standard operating procedures or instructions of the radiologist in charge.
- Identify each patient, review requisition, and record pertinent data on requisition at all times.
- Assist radiologists and other physicians in performing examinations.
- Review each examination radiograph for technical accuracy.
- Capture the correct anatomical part on all required images, and ensure proper film identification on all images.
- Ensure adherence to safety standards.
- Communicate all pertinent data to persons responsible for the care of the patient following the procedure.
- Report malfunction of equipment to your supervisor.
- Assess patient condition; ensure patient safety through proper use of restraints/support devices.
- Maintain cleanliness of equipment and work area.
- Maintain daily patient log.
- Train and assist staff and students.
- Direct technologists and students in the proper use of equipment.
- Assist technologists and students in production of quality images.
- Perform data entry into the radiology information system and hospital information system.
- Maintain inventory of supplies.
- Maintain confidentiality of all hospital or imaging center information.
- Obtain a minimum of 24 category A continuing education credits every two years (on your own biennium schedule), as per ARRT requirements.

Secondary Duties and Responsibilities

- Request supplies as needed.
- Perform associated clerical functions.
- Perform clearly related duties, as directed.
- Perform clearly unrelated duties, as directed in an emergency setting.

General Qualities and Duties

- Sensitive to the needs of others (empathy)
- Friendly and helpful
- Demonstrates judgment and tact
- Cooperative
- Communicates effectively
- Performs duties in compliance with all appropriate safety and infection control guidelines
- Complies with all guidelines, including all, but not limited to, attendance and lateness guidelines
- Complies with dress code
- Demonstrates initiative
- Demonstrates an understanding of the hospital's mission statement or the corporation's goals and standards.

Continuing Education

What is continuing education? Why do I need credits? What type of credits do I need? How many credits do I need to renew my certification/license? Do I need credits if I just passed my registry? How often do I need to get continuing education credits?

What is continuing education? Continuing education, also commonly referred to as secondary education, is used to either obtain additional certifications or credits required to maintain licensure. Continuing education can be taken in the form of workshops, seminars, home study, online courses, conferences, or hands-on training. Mandatory continuing education is common for health professionals and is required by law to keep up with ever-changing technology, thus guaranteeing the best possible end results for your patients.

Why do I need credits? As of 1995, the American Registry of Radiologic Technologists mandated that all technologists obtain continuing education credits to renew or reinstate certification. Many states have recognized the ARRT as the foundation for radiologic technologists and accept CE credits submitted for your biennium as sufficient for state renewal.

What type of credits do I need? All continuing education credits must be Category A or A+ for renewal purposes. As of 2008, Category B credits are no longer accepted for submission. It would be best to consult with the course instructors or facilitators prior to enrollment to ensure that all course materials have been approved for ARRT submission.

How many credits do I need to renew my certification/license? The ARRT requires 24 credits biannually to renew certification. There are several ways to earn credits that will satisfy your biennium requirements:

- Earn 24 credits set forth by the guidelines of the ARRT, such as attending seminars, self-studies, workshops, online courses, and so on. These courses must be approved by the American Society of Radiologic Technologists (ASRT) to be recognized by the ARRT.
- Successfully pass a primary examination in a category not previously passed. Here are some examples of primary examinations: **Diagnostic Radiography, Nuclear Medicine Technology, Radiation Therapy, Magnetic Resonance Imaging, Dosimetry, Diagnostic Medical Sonography, Radiology Administration, Sonography, Vascular Technology, Diagnostic Cardiac Sonography.**
- Successfully pass a postprimary examination in a category not previously passed. Here are some examples of postprimary examinations: **Mammography, Computed Tomography, Magnetic Resonance Imaging, Quality Management, Sonography, Vascular Sonography, Bone Densitometry, Vascular-Interventional**

Radiography, Cardiac-Interventional Radiography, Breast Sonography, Radiologist Assistant, Nuclear Cardiology, Positron Emission Tomography, Certified Imaging Informatics Professional.

Do I need credits if I just passed my registry? No. Upon successfully passing the ARRT radiography registry, you automatically receive 24 credits, which will be applied toward your biennium. However, you are required to begin the fulfillment of your CE requirements on the first day of the next birth month after the examination. For example, if you successfully pass the registry in January 2011, but are born in May, you will need to begin your fulfillment of 24 credits on June 1, 2011 for your 2013 renewal.

How often do I need to get continuing education credits? The continuing education requirements are based on a two-year cycle in relation to the registrant's birth month. The biennium will begin on the first day of your birth month and expire on the last day of the month prior. For example, if your birth month is February, your registration is valid February 1, 2011 through January 31, 2013.

2 ▶ CERTIFICATION EXAM IN RADIOGRAPHY

CHAPTER SUMMARY

This section will give you a better understanding of how the registry examination is administered by the American Registry of Radiologic Technologists. We will discuss how to apply for the registry, test format, categories, and scoring.

The American Registry of Radiologic Technologists administers the radiography exam for certification purposes only. This test measures the candidate's understanding and knowledge of the responsibilities that an entry-level staff technologist would generally need to possess to succeed in the field of diagnostic radiography. All candidates must meet specific eligibility requirements prior to sitting for the examination.

About the AART

How to Apply

If you are a newly graduated student, your program director should have supplied you with a handbook provided to them by the ARRT. This handbook contains all of the appropriate materials, information, and documents that you will need to submit to be registry eligible. If you have not received a handbook from your program director or you are not a newly graduated student, you may contact the American Registry of Radiologic Technologists directly to have your handbook mailed to your home. Registry-eligible candidates must have—within the past five years—successfully completed a formal educational program that is recognized and accredited by a mechanism of the ARRT. Candidates will be given three attempts to successfully pass the examination. Following are a few items that you will need to do to be considered eligible to take the ARRT:

- provide a passport-quality photograph of yourself
- complete the application in detail
- receive endorsement of application by program director or authorized faculty member
- submit the completed application along with payment to ARRT
- supply any additional information requested by ARRT to determine eligibility

Test Format, Categories, and Scoring

All individuals will be given 3.5 hours to finish an electronic 220-question multiple-choice exam. Your final score will be based on only 200 questions. There are 20 pilot questions that have been randomly placed in the examination and they will not be recognizable to the test taker. These purpose of these pilot questions is for the ARRT to determine which new questions and content can be used in future registries. The raw score (the number answered correctly) will be converted to a scaled score in the 0–99 range. The minimum passing scaled score is 75%. Following is a breakdown of what you can expect to find on the test:

Radiation Protection, 20%, 40 Questions
- Biological aspects of radiation
- Minimizing patient exposure
- Personnel protection
- Radiation exposure and monitoring

Equipment Operation and Quality Control, 12%, 24 Questions
- Principles of radiation physics
- Radiographic equipment
- Quality control of radiographic equipment and accessories

Image Production and Evaluation, 25%, 50 Questions
- Selection of technical factors
- Image processing and quality assurance
- Criteria for image evaluation

Radiographic Procedures, 30%, 60 Questions
- Thorax
- Abdomen and gi studies
- Urological studies
- Spine and pelvis
- Cranium
- Extremities
- Other

Patient Care and Education, 13%, 26 Questions
- Ethical and legal aspects
- Interpersonal communication
- Infection control
- Physical assistance and transfer
- Medical emergencies
- Contrast media

3 ▶ THE LEARNINGEXPRESS TEST PREPARATION SYSTEM

CHAPTER SUMMARY

Taking any written exam can be tough and a top score demands a lot of preparation. The LearningExpress Test Preparation System, developed exclusively for LearningExpress by leading test experts, gives you the discipline and attitude you need to be a winner.

Taking this written exam is no picnic, and neither is getting ready for it. Your future career in radiography depends on whether you pass the test, and there are all sorts of pitfalls that can keep you from doing your best on this all-important exam. Here are some of the obstacles that can stand in the way of your success:

- Being unfamiliar with the format of the exam
- Being paralyzed by test anxiety
- Leaving your preparation to the last minute or not preparing at all
- Not knowing vital test-taking skills: how to pace yourself through the exam, how to use the process of elimination, and when to guess
- Not being in tip-top mental and physical shape
- Messing up on exam day by having to work on an empty stomach or shivering through the exam because the room is cold

What's the common denominator in all of these test-taking pitfalls? One word: *control*. Who's in control, you or the exam?

LearningExpress Test Preparation System

The LearningExpress Test Preparation System puts you in control. In just nine easy-to-follow steps, you will learn everything you need to know to make sure that you are in charge of your preparation and your performance on the exam. Other test takers may let the exam get the better of them; other test takers may be unprepared or out of shape, but not you. After completing this chapter, you will have taken all the steps you need to get a high score on the radiologic technologist exam.

Here's how the LearningExpress Test Preparation System works: Each of the following nine steps includes one or more activities. It's important that you do the activities along with the reading, or you won't be getting the full benefit of the system. Each step tells you approximately how much time you should spend.

> **Step 1:** Get Information (30 minutes)
> **Step 2:** Conquer Test Anxiety (20 minutes)
> **Step 3:** Make a Plan (50 minutes)
> **Step 4:** Learn to Manage Your Time (10 minutes)
> **Step 5:** Learn to Use the Process of Elimination (20 minutes)
> **Step 6:** Know When to Guess (20 minutes)
> **Step 7:** Reach Your Peak Performance Zone (10 minutes)
> **Step 8:** Get Your Act Together (10 minutes)
> **Step 9:** Do It! (10 minutes)
>
> **Total time for complete system:**
> 180 minutes—three hours

We estimate that working through the entire system will take you approximately three hours. It's perfectly okay if you work at a faster or slower pace. If you can spend a whole afternoon or evening working, you can work through the whole LearningExpress Test Preparation System in one sitting. Otherwise, you can break it up, and do just one or two steps a day for the next several days. It's up to you—remember, you are in control.

Step 1: Get Information

Time to complete: 30 minutes
Activities: Read Chapter 1, Overview of a Career as a Radiographer

Knowledge is power. The first step in the LearningExpress Test Preparation System is finding out everything you can about the radiologic technologist exam. The more information you have, the more in control you will feel.

What You Should Find Out

The more details you know about the exam, the more efficiently you will be able to study. Here's a list of some things you might want to find out about your exam:

- Will your test be computer-based or use a paper booklet and answer sheet?
- What skills are tested?
- How many sections are on the exam?
- How many questions are in each section?
- Are the questions ordered from easy to hard, or is the sequence random?
- How much time is allotted for each section?
- Are there breaks between sections?
- What is the passing score, and how many questions do you have to answer correctly in order to get that score?
- Does a higher score give you any advantages?

- How is the exam scored, and is there a penalty for incorrect answers?
- Are you permitted to go back to a prior section or move on to the next section if you finish early?
- Can you write in the exam booklet for a paper test? Will you be given scratch paper?
- What should you bring with you on exam day?

Step 2: Conquer Test Anxiety

Time to complete: 20 minutes

Activity: Take the Test Anxiety Quiz (page 11)

Having complete information about the exam is the first step in getting control of it. Next, you have to overcome one of the biggest obstacles to test success: test anxiety. Test anxiety can not only impair your performance on the exam itself, but can even keep you from preparing properly. In Step 2, you will learn stress management techniques that will help you succeed on your exam. Learn these strategies now, and practice them as you work through the questions in this book so they'll be second nature to you by exam day.

Combating Test Anxiety

The first thing you need to know is that a little test anxiety is a good thing. Everyone gets nervous before a big exam—and if that nervousness motivates you to prepare thoroughly, so much the better. It's said that Sir Laurence Olivier, one of the foremost British actors of the twentieth century, was ill before every performance. And younger actors—including Zac Efron and Claire Danes—also get the jitters. Their stage fright doesn't impair their performances, though; in fact, it probably gives them a little extra edge—just the kind of edge you need to do well, whether on a stage, in a film, or in an examination room.

The Test Anxiety Quiz is on page 11. Stop now and answer the questions on that page to find out whether your level of test anxiety is something you should worry about.

Stress Management before the Exam

If you feel your level of anxiety is getting the best of you in the weeks before the exam, here is what you need to do to bring the level down again:

- **Get prepared.** There's nothing like knowing what to expect and being prepared for it to put you in control of test anxiety. That's why you're reading this book. Use it faithfully, and remind yourself that you're better prepared than most of the people taking the exam.
- **Practice self-confidence.** A positive attitude is a great way to combat test anxiety. This is no time to be humble or shy. Stand in front of the mirror and say to your reflection, "I'm prepared. I'm full of self-confidence. I'm going to ace this exam. I know I can do it." Say it into a recorder, and play it back once a day. Or type it into a text message and send it to yourself. It may feel a little weird, but if you hear it or see it often enough, you will believe it.
- **Fight negative messages.** Every time someone starts telling you how hard the exam is or how it's almost impossible to pass the test, start telling that person the self-confidence messages you've been practicing. If the person delivering the negative messages is you—telling yourself that you don't do well on exams, that you just can't do it—don't listen. Turn on your recorder, or reread your text messages, and listen to yourself expressing confidence in your own abilities.
- **Visualize.** Imagine yourself reporting for duty on your first day working as a radiologic technologist. Think of yourself wearing your uniform with pride and learning skills you will use for the rest of your professional life. Visualizing success can help make it happen—and remind you of why you're doing all this work preparing for the exam.
- **Exercise.** Physical activity helps calm down your body and focus your mind. Besides, being in good physical shape can actually help you do well on the exam. Go for a run, lift weights, go swimming—and do it regularly.

Stress Management on Exam Day

There are several ways you can bring down your level of test stress and anxiety on exam day. They'll work best if you practice them in the weeks before the exam, so you know which ones work best for you.

- **Deep breathing.** Take a deep breath while you count to five. Hold it for a count of one, and then let it out on a count of five. Repeat several times.
- **Move your body.** Try rolling your head in a circle. Rotate your shoulders. Shake your hands from the wrist. Many people find these movements very relaxing.
- **Visualize again.** Think of the place where you are most relaxed: lying on the beach in the sun, walking through the park, or whatever relaxes you. Now, close your eyes and imagine you're there. If you practice in advance, you will find that you need only a few seconds of this exercise to experience a significant increase in your sense of well-being.

When anxiety threatens to overwhelm you during the exam, there are still things you can do to manage your stress level:

- **Repeat your self-confidence messages.** You should have them memorized by now. Say them quietly to yourself, and believe them!
- **Visualize one more time.** This time, visualize yourself moving smoothly and quickly through the exam, answering every question correctly and finishing just before time is up. Like most visualization techniques, this one works best if you've practiced it ahead of time.
- **Find an easy question.** If the test format allows, skim over the test until you find an easy question, and answer it. Completing even one question gets you into the test-taking groove.
- **Take a mental break.** Everyone loses concentration once in a while during a long exam. It's normal, so you shouldn't worry about it. Instead, accept what has happened. Say to yourself, "Hey, I lost it there for a minute. My brain is taking a break." Put down your pencil, close your eyes, and do some deep breathing for a few seconds. Then, you're ready to go back to work.

Try these techniques ahead of time, and see if they work for you!

Step 3: Make a Plan

Time to complete: 50 minutes

Activity: Construct a study plan, using Schedules A–B (pages 12–13)

Many people do poorly on exams because they forget to make a study schedule. The most important thing you can do to better prepare yourself for your exam is to create a study plan or schedule. Spending hours the day before the exam poring over sample test questions not only raises your level of anxiety, but also is not a substitute for careful preparation and practice over time.

Don't cram. Take control of your time by mapping out a study schedule. There are two examples of study schedules on pages 12–13, based on the amount of time you have before the exam. If you're the kind of person who needs deadlines and assignments to motivate you for a project, here they are. If you're the kind of person who doesn't like to follow other people's plans, you can use the suggested schedules to construct your own.

Even more important than making a plan is making a commitment. You can't review everything you learned in your radiologic technologist program in one night. You have to set aside some time every day for studying and practice. Aim for at least 20 minutes a day. This will do you more good than two hours crammed into a Saturday.

If you have months before the exam, you're lucky. Don't put off your study until the week before the exam. Start now. Even ten minutes a day, with half an hour or more on weekends, can make a big difference in your score—and in your chances of becoming a radiologic technologist.

Test Anxiety Quiz

You need to worry about test anxiety only if it is extreme enough to impair your performance. The following question-naire will provide a diagnosis of your level of test anxiety. In the blank before each statement, write the number that most accurately describes your experience.

0 = Never   1 = Once or twice   2 = Sometimes   3 = Often

_____ I have gotten so nervous before an exam that I simply put down the books and didn't study for it.

_____ I have experienced disabling physical symptoms such as vomiting and severe headaches because I was nervous about an exam.

_____ I have simply not shown up for an exam because I was scared to take it.

_____ I have experienced dizziness and disorientation while taking an exam.

_____ I have had trouble filling in the little circles because my hands were shaking too hard.

_____ I have failed an exam because I was too nervous to complete it.

_____ **Total: Add up the numbers in the blanks above.**

Your Test Stress Score

Here are the steps you should take, depending on your score. If you scored:

- **Below 3,** your level of test anxiety is nothing to worry about; it's probably just enough to give you that little extra edge.
- **Between 3 and 6,** your test anxiety may be enough to impair your performance, and you should practice the stress management techniques listed in this chapter to try to bring your test anxiety down to manageable levels.
- **Above 6,** your level of test anxiety is a serious concern. In addition to practicing the stress management techniques listed in this chapter, you may want to seek additional, personal help. Call your local high school or community college and ask for the academic counselor. Tell the counselor that you have a level of test anxiety that sometimes keeps you from being able to take the exam. The counselor may be willing to help you or may suggest someone else you should talk to.

Schedule A: Leisure Plan

This schedule gives you at least five months to sharpen your skills and prepare for your exam. The more prep time you give yourself, the more relaxed you'll feel.

TIME	PREPARATION
5 months before the test	Read the Introduction and Chapters 1 and 2. Start going to the library once every two weeks to read books or magazines about radiologic technologists. Start gathering information about working as a radiologic technologist. Find other people who are preparing for the exam, and form a study group with them.
4 months before the test	Read Chapter 3 and work through the practice questions.
3 months before the test	Read Chapter 4 and work through the practice questions. You're still continuing with your reading, aren't you?
2 months before the test	Read Chapter 5 and work through the practice questions.
1 month before the test	Use your scores from the practice questions in Chapter 3, 4, and 5 to help you decide where to concentrate your efforts this month. Go back to the relevant chapters and review challenging topics. Continue working with your study group.
1 week before the test	Take and review the sample exams in Chapters 6 and 7. See how much you've learned in the past months. Concentrate on what you've done well, and decide not to let any areas where you still feel uncertain bother you.
1 day before the test	Relax. Do something unrelated to the radiologic technologist exam. Eat a good meal and go to bed at your usual time.

Schedule B: The Just-Enough-Time Plan

If you have three to four months before your exam, that should be enough time to prepare. This schedule assumes three months; stretch it out or compress it if you have more or less time.

TIME	PREPARATION
5 months before the test	Read the Introduction and Chapters 1 and 2. Start going to the library once every two weeks to read books or magazines about radiologic technologists. Start gathering information about working as a radiologic technologist. Find other people who are preparing for the exam, and form a study group with them.
2 months before the test	Read Chapters 3, 4, and 5, and work through the exercises.
1 months before the test	Take one of the sample exams in Chapters 6 and 7. Use your score to help you decide where to concentrate your efforts this month. Go back to the relevant chapters and review challenging topics, or get the help of a friend or teacher.
1 week before the test	Review the sample exams in Chapters 6 and 7. See how much you have learned in the past months. Celebrate what you've done well, and decide not to let any areas where you still feel uncertain bother you.
1 day before the test	Relax. Do something unrelated to the radiologic technologist exam. Eat a good meal and go to bed at your usual time.

Step 4: Learn to Manage Your Time

Time to complete: 10 minutes to read, many hours of practice

Activities: Practice these strategies as you take the sample exams

Steps 4, 5, and 6 of the LearningExpress Test Preparation System put you in charge of your exam by showing you test-taking strategies that work. Practice these strategies as you take the sample exams in Chapters 6 and 7. Then, you will be ready to use them on exam day.

First, you will take control of your time on the exam. The first step in achieving this control is to find out the format of the exam you're going to take. The radiologic technologist exam covers five different content areas, which you will become familiar with. You will want to practice using your time wisely on the practice exams and trying to avoid mistakes while working quickly. Practice pacing yourself on the practice exams so you don't spend too much time on difficult questions. Here are some more tips:

- **Listen carefully to directions.** By the time you get to the exam, you should know how the test works, but listen just in case something has changed.
- **Pace yourself.** Glance at your watch every few minutes, and compare the time to how far you've gotten in the section. Leave some extra time for review so that when one-quarter of the time has elapsed, you should be more than one-quarter of the way through the section, and so on. If you're falling behind, pick up the pace.
- **Keep moving.** Don't spend too much time on one question. If you don't know the answer, skip the question and move on. Circle the number of the question in your test booklet in case you have time to come back to it later.
- **Keep track of your place on the answer sheet.** If you skip a question, make sure you skip it on the answer sheet, too. Check yourself every five to ten questions to make sure the question number and the answer sheet number match.
- **Don't rush.** You should keep moving; rushing won't help. Try to keep calm and work methodically and quickly.

Step 5: Learn to Use the Process of Elimination

Time to complete: 20 minutes

Activity: Complete the worksheet on Using the Process of Elimination (page 16)

After time management, the next most important tool for taking control of your exam is using the process of elimination wisely. It's standard test-taking wisdom that you should always read all the answer choices before choosing your answer. This helps you find the correct answer by eliminating incorrect answer choices. And, sure enough, that standard wisdom applies to this exam, too.

Let's say you are facing a question that goes like this:

"Biology uses a binomial system of classification." In this sentence, the word *binomial* most nearly means
a. understanding the law.
b. having two names.
c. scientifically sound.
d. having a double meaning.

If you happen to know what *binomial* means, of course you don't need to use the process of elimination, but let's assume you don't. So, you look at the answer choices. *Understanding the law* sure doesn't sound very likely for something having to do with biology. So you eliminate choice **a**—and now you have only three answer choices to deal with.

Now, move on to the other answer choices. If you know that the prefix *bi-* means *two,* as in *bicycle,* you would consider choice **b** as a possible answer.

Choice **c**, *scientifically sound*, is a possibility. At least it's about science, not law. It could work here;

though, when you think about it, having a scientifically sound classification system in a scientific field is kind of redundant. You remember the *bi-* in *binomial,* and probably continue to like choice **b** better. But you're not sure, so **c** is a maybe.

Now, let's look at choice **d**, *having a double meaning.* You're still keeping in mind that *bi-* means *two,* so this one looks possible at first. But then you look again at the sentence the word is in, and you think, "Why would biology want a system of classification that has two meanings? That wouldn't work very well!" If you're really taken with the idea that *bi-* means *two,* you might put a question mark here. But if you're feeling a little more confident, you'll eliminate **d** because you already have a better answer picked out.

Now your question looks like this:

"Biology uses a binomial system of classification." In this sentence, the word *binomial* most nearly means
  ✗ **a.** understanding the law.
  ✓ **b.** having two names.
  ? **c.** scientifically sound.
  ? **d.** having a double meaning.

You've got just one check mark for a good answer. If you're pressed for time, you should simply mark choice **b** on your answer sheet. If you have the time to be extra careful, you could compare your check-mark answer to your question-mark answers to make sure that it's better. (It is: The *binomial* system in biology is the one that gives a two-part genus and species name, like *Homo sapiens.*)

Even when you think you are absolutely clueless about a question, you can often use the process of elimination to get rid of at least one answer choice. If you do this, you will be better prepared to make an educated guess, as you will see in Step 6. Often, you can eliminate answers until you have only two possible answers. Then you are in a strong position to guess.

Try using your elimination skills on the questions in the worksheet on page 16, Using the Process of Elimination. The questions are not about radiography; they are just designed to show you how the process of elimination works. The answer explanations for this worksheet show one possible way you might use the process to arrive at the right answer.

The process of elimination is your tool for the next step, which is knowing when to guess.

Using the Process of Elimination

Use the process of elimination to answer the following questions.

1. Ilsa is as old as Meghan will be in five years. The difference between Ed's age and Meghan's age is twice the difference between Ilsa's age and Meghan's age. Ed is 29. How old is Ilsa?
   a. 4
   b. 10
   c. 19
   d. 24

2. "All drivers of commercial vehicles must carry a valid commercial driver's license whenever operating a commercial vehicle." According to this sentence, which of the following people need NOT carry a commercial driver's license?
   a. a truck driver idling his engine while waiting to be directed to a loading dock
   b. a bus operator backing her bus out of the way of another bus in the bus lot
   c. a taxi driver driving his personal car to the grocery store
   d. a limousine driver taking the limousine to her home after dropping off her last passenger of the evening

3. Smoking tobacco has been linked to
   a. increased risk of stroke and heart attack.
   b. all forms of respiratory disease.
   c. increasing mortality rates over the past ten years.
   d. juvenile delinquency.

4. Which of the following words is spelled correctly?
   a. incorrigible
   b. outragous
   c. domestickated
   d. understandible

Answers

Here are the answers, as well as some suggestions as to how you might have used the process of elimination to find them.

1. **d.** You should have eliminated choice **a** off the bat. Ilsa can't be four years old if Meghan is going to be Ilsa's age in five years. The best way to eliminate other answer choices is to try plugging them in to the information given in the problem. For instance, for choice **b**, if Ilsa is 10, then Meghan must be 5. The difference in their ages is 5. The difference between Ed's age, 29, and Meghan's age, 5, is 24. Is 24 two times 5? No. Then choice **b** is wrong. You could eliminate choice **c** in the same way and be left with choice **d**.

2. **c.** Note the word *not* in the question, and go through the answers one by one. Is the truck driver in choice **a** "operating a commercial vehicle"? Yes, idling counts as "operating," so he needs to have a commercial driver's license. Likewise, the bus operator in choice **b** is operating a commercial vehicle; the question doesn't say the operator has to be on the street. The limo driver in choice **d** is operating a commercial vehicle, even if it doesn't have a passenger in it. However, the cabbie in choice **c** is not operating a commercial vehicle, but his own private car.

3. a. You could eliminate choice **b** simply because of the presence of the word *all*. Such absolutes hardly ever appear in correct answer choices. Choice **c** looks attractive until you think a little about what you know—aren't fewer people smoking these days, rather than more? So how could smoking be responsible for a higher mortality rate? (If you didn't know that *mortality rate* means the rate at which people die, you might keep this choice as a possibility, but you would still be able to eliminate two answers and have only two to choose from.) And choice **d** is unlikely, so you could eliminate that one, too. You are left with the correct choice, **a**.

4. a. How you used the process of elimination here depends on which words you recognized as being spelled incorrectly. If you knew that the correct spellings were outrageous, domesticated, and understandable, then you were home free.

Step 6: Know When to Guess

Time to complete: 20 minutes

Activity: Complete worksheet on Your Guessing Ability (pages 19–20)

Armed with the process of elimination, you're ready to take control of one of the big questions in test taking: Should I guess?

The main answer is yes. The radiologic technologist exams don't use a so-called guessing penalty. Basically, the number of questions you answer correctly yields your score, and there's no penalty for wrong answers. So, you don't have to worry—simply go ahead and guess.

How well you will do with guessing depends on you, your personality, and your "guessing intuition."

When There Is No Guessing Penalty

As previously noted, the radiologic technologist exams don't have a guessing penalty. That means, all other things being equal, you should always go ahead and guess, even if you have no idea what the question means. Nothing can happen to you if you're wrong. But all other things aren't necessarily equal. The other factor in deciding whether or not to guess, besides the guessing penalty, is you. There are two things you need to know about yourself before you go into the exam:

- Are you a risk taker?
- Are you a good guesser?

Your risk-taking temperament matters most on exams with a guessing penalty. Without a guessing penalty, even if you're a play-it-safe person, guessing is perfectly safe. Overcome your anxieties, and go ahead and mark an answer.

But what if you're not much of a risk taker, and you think of yourself as the world's worst guesser? Complete the Your Guessing Ability worksheet on pages 19–20 to get an idea of how good your intuition is.

Step 7: Reach Your Peak Performance Zone

Time to complete: 10 minutes to read, weeks to complete!

Activity: Exercise, proper diet, and rest

To get ready for a challenge like a big exam, you not only have to take control of your mental state, but also your physical state. Exercise, proper diet, and rest will ensure that your body works with, rather than against, your mind to prepare for test day.

Exercise

If you don't already have a regular exercise program going, the time during which you're preparing for an exam is an excellent time to start one. And if you're already trying to keep fit, don't let the pressure of preparing for an exam fool you into quitting now. Exercise helps reduce stress by pumping wonderful good-feeling hormones called endorphins into your system. It also increases the oxygen supply throughout your body, including to your brain, so you will be at peak performance on exam day.

A half hour of vigorous activity—enough to break a sweat—every day should be your aim. If you're really pressed for time, every other day is okay. Choose an activity you like and get out there and do it. Jogging with a friend always makes the time go faster, as does running with a radio or MP3 player.

But don't overdo it. You don't want to exhaust yourself. Moderation is the key.

Diet

First, cut out the junk. Go easy on caffeine, and try to eliminate alcohol and nicotine from your system at least two weeks before the exam.

What your body needs for peak performance is simply a balanced diet. Eat plenty of fruits and vegetables, along with protein and complex carbohydrates. Foods that are high in lecithin (an amino acid), such as fish and beans, are especially good "brain foods."

The night before the exam, you might load up on carbohydrates the way athletes do before a competition. Eat a big plate of spaghetti, rice and beans, or any of your favorite carbohydrates.

Rest

You probably know how much sleep you need every night to be at your best, even if you don't always get it. Make sure you do get that much sleep, though, for at least a week before the exam. Moderation is important here, too. Too much sleep will just make you groggy.

If you're not a morning person and your exam will be given in the morning, you should reset your internal clock so that your body doesn't think you're taking an exam at 3 a.m. You have to start this process well before the exam. The way it works is to get up half an hour earlier each morning, and then go to bed half an hour earlier that night. Don't try it the other way around; you will just toss and turn if you go to bed early without having gotten up early. The next morning, get up another half an hour earlier, and so on. How long you will have to do this depends on how late you're used to getting up.

Step 8: Get Your Act Together

Time to complete: 10 minutes to read; time to complete will vary

Activity: Know your exam site and the materials you'll need

You're in control of your mind and body; you're in charge of your test anxiety, your preparation, and your test-taking strategies. Now, it's time to take charge of external factors, like the exam site and the materials you need to take the exam.

Find Out Where the Exam Is and Make a Trial Run

The testing agency or your radiologic technologist exam instructor will notify you when and where your exam is being held. Do you know how to get to the exam site? Do you know how long it will take to get there? If not, make a trial run, preferably on the same day of the week at the same time of day. Plan to arrive 10 to 15 minutes early so you can get the lay of the land, use the bathroom, and calm down. Then, figure out how early you will have to get up that morning, and make sure you get up that early every day for a week before the exam.

Gather Your Materials

The night before the exam, lay out the clothes you will wear and the materials you have to bring with you to the exam. Plan on dressing in layers; you won't have any control over the temperature of the examination room. Have a sweater or jacket you can take off if it's warm. Use the checklist on the Final Preparations worksheet on page 22 to help you pull together what you will need.

Your Guessing Ability

The following are ten really hard questions. You are not supposed to know the answers. Rather, this is an assessment of your ability to guess when you don't have a clue. Read each question carefully, just as if you did expect to answer it. If you have any knowledge of the subject, use that knowledge to help you eliminate wrong answer choices.

1. September 7 is Independence Day in
   a. India.
   b. Costa Rica.
   c. Brazil.
   d. Australia.

2. Which of the following is the formula for determining the momentum of an object?
   a. $p = MV$
   b. $F = ma$
   c. $P = IV$
   d. $E = mc^2$

3. Because of the expansion of the universe, the stars and other celestial bodies are all moving away from each other. This phenomenon is known as
   a. Newton's first law.
   b. the big bang.
   c. gravitational collapse.
   d. Hubble flow.

4. American author Gertrude Stein was born in
   a. 1713.
   b. 1830.
   c. 1874.
   d. 1901.

5. Which of the following is NOT one of the Five Classics attributed to Confucius?
   a. *I Ching*
   b. *Book of Holiness*
   c. *Spring and Autumn Annals*
   d. *Book of History*

6. The religious and philosophical doctrine that holds that the universe is constantly in a struggle between good and evil is known as
   a. Pelagianism.
   b. Manichaeanism.
   c. neo-Hegelianism.
   d. Epicureanism.

7. The third chief justice of the U.S. Supreme Court was
   a. John Blair.
   b. William Cushing.
   c. James Wilson.
   d. John Jay.

8. Which of the following is the poisonous portion of a daffodil?
   a. the bulb
   b. the leaves
   c. the stem
   d. the flowers

9. The winner of the Masters golf tournament in 1953 was

  a. Sam Snead.

  b. Cary Middlecoff.

  c. Arnold Palmer.

  d. Ben Hogan.

10. The state with the highest per capita personal income in 1980 was

  a. Alaska.

  b. Connecticut.

  c. New York.

  d. Texas.

Answers

Check your answers against the correct answers below.

  1. c.

  2. a.

  3. d.

  4. c.

  5. b.

  6. b.

  7. b.

  8. a.

  9. d.

  10. a.

How Did You Do?

You may have simply gotten lucky and actually known the answers to one or two questions. In addition, your guessing was probably more successful if you were able to use the process of elimination on any of the questions. Maybe you didn't know who the third chief justice was (question 7), but you knew that John Jay was the first. In that case, you would have eliminated choice **d** and therefore improved your odds of guessing right from one in four to one in three.

According to probability, you should get two and a half answers correct, so getting either two or three right would be average. If you got four or more right, you may be a really terrific guesser. If you got one or none right, you may be a poor guesser.

Keep in mind, though, that this is only a small sample. You should continue to keep track of your guessing ability as you work through the sample questions in this book. Circle the numbers of questions you guess on as you make your guess; or, if you don't have time while you take the practice tests, go back afterward and try to remember which questions you guessed at.

Remember, on a test with four answer choices, your chance of guessing correctly is one in four. So keep a separate "guessing score" for each exam. How many questions did you guess on? How many did you get right? If the number you got right is at least one-fourth of the number of questions you guessed on, you are at least an average guesser—maybe better—and you should always go ahead and guess on the real exam. If the number you got right is significantly lower than one-fourth of the number you guessed on, you would be safe in guessing anyway. Because the radiologic technologist exams have no guessing penalty, even if you are a play-it-safe person with lousy intuition, you are still safe guessing every time.

Don't Skip Breakfast

Even if you don't usually eat breakfast, do so on exam morning. A cup of coffee doesn't count. Don't choose doughnuts or other sweet foods, either. A sugar high will leave you with a sugar low in the middle of the exam. A mix of protein and carbohydrates is best. Cereal with milk or eggs with toast will do your body a world of good.

Step 9: Do It!

Time to complete: 10 minutes, plus test-taking time
Activity: Ace the radiologic technologist exam!
Fast-forward to exam day. You're ready. You made a study plan and followed through. You practiced your test-taking strategies while working through this book. You're in control of your physical, mental, and emo-

tional state. You know when and where to show up and what to bring with you. In other words, you're better prepared than most of the other people taking the radiologic technologist exam with you. You're psyched.

Just one more thing. When you're done with the exam, you will have earned a reward. Plan a celebration. Call up your friends and plan a party, or have a nice dinner for two, or pick out a movie to see—whatever your heart desires. Give yourself something to look forward to.

And then do it. Go into the exam, full of confidence, armed with test-taking strategies you've practiced until they're second nature. You're in control of yourself, your environment, and your performance on the exam. You're ready to succeed. So do it. Go in there and ace the exam. And look forward to your future career as a radiologic technologist.

Getting to the Exam Site

Location of exam site: _____

Date: _____

Departure time: _____

Do I know how to get to the exam site?   Yes ____   No ____   (If no, make a trial run.)

Time it will take to get to exam site: _____

Things to Lay out the Night before

Clothes I will wear   ____

Sweater/jacket   ____

Watch   ____

Photo ID   ____

Four #2 pencils (for paper-based exam)   ____

Other Things to Bring/Remember

_____   _____

_____   _____

_____   _____

_____   _____

CHAPTER

4 ▶ RADIOGRAPHY PRACTICE TEST I

Radiography Practice Test I

1.	(a)	(b)	(c)	(d)		37.	(a)	(b)	(c)	(d)		73.	(a)	(b)	(c)	(d)
2.	(a)	(b)	(c)	(d)		38.	(a)	(b)	(c)	(d)		74.	(a)	(b)	(c)	(d)
3.	(a)	(b)	(c)	(d)		39.	(a)	(b)	(c)	(d)		75.	(a)	(b)	(c)	(d)
4.	(a)	(b)	(c)	(d)		40.	(a)	(b)	(c)	(d)		76.	(a)	(b)	(c)	(d)
5.	(a)	(b)	(c)	(d)		41.	(a)	(b)	(c)	(d)		77.	(a)	(b)	(c)	(d)
6.	(a)	(b)	(c)	(d)		42.	(a)	(b)	(c)	(d)		78.	(a)	(b)	(c)	(d)
7.	(a)	(b)	(c)	(d)		43.	(a)	(b)	(c)	(d)		79.	(a)	(b)	(c)	(d)
8.	(a)	(b)	(c)	(d)		44.	(a)	(b)	(c)	(d)		80.	(a)	(b)	(c)	(d)
9.	(a)	(b)	(c)	(d)		45.	(a)	(b)	(c)	(d)		81.	(a)	(b)	(c)	(d)
10.	(a)	(b)	(c)	(d)		46.	(a)	(b)	(c)	(d)		82.	(a)	(b)	(c)	(d)
11.	(a)	(b)	(c)	(d)		47.	(a)	(b)	(c)	(d)		83.	(a)	(b)	(c)	(d)
12.	(a)	(b)	(c)	(d)		48.	(a)	(b)	(c)	(d)		84.	(a)	(b)	(c)	(d)
13.	(a)	(b)	(c)	(d)		49.	(a)	(b)	(c)	(d)		85.	(a)	(b)	(c)	(d)
14.	(a)	(b)	(c)	(d)		50.	(a)	(b)	(c)	(d)		86.	(a)	(b)	(c)	(d)
15.	(a)	(b)	(c)	(d)		51.	(a)	(b)	(c)	(d)		87.	(a)	(b)	(c)	(d)
16.	(a)	(b)	(c)	(d)		52.	(a)	(b)	(c)	(d)		88.	(a)	(b)	(c)	(d)
17.	(a)	(b)	(c)	(d)		53.	(a)	(b)	(c)	(d)		89.	(a)	(b)	(c)	(d)
18.	(a)	(b)	(c)	(d)		54.	(a)	(b)	(c)	(d)		90.	(a)	(b)	(c)	(d)
19.	(a)	(b)	(c)	(d)		55.	(a)	(b)	(c)	(d)		91.	(a)	(b)	(c)	(d)
20.	(a)	(b)	(c)	(d)		56.	(a)	(b)	(c)	(d)		92.	(a)	(b)	(c)	(d)
21.	(a)	(b)	(c)	(d)		57.	(a)	(b)	(c)	(d)		93.	(a)	(b)	(c)	(d)
22.	(a)	(b)	(c)	(d)		58.	(a)	(b)	(c)	(d)		94.	(a)	(b)	(c)	(d)
23.	(a)	(b)	(c)	(d)		59.	(a)	(b)	(c)	(d)		95.	(a)	(b)	(c)	(d)
24.	(a)	(b)	(c)	(d)		60.	(a)	(b)	(c)	(d)		96.	(a)	(b)	(c)	(d)
25.	(a)	(b)	(c)	(d)		61.	(a)	(b)	(c)	(d)		97.	(a)	(b)	(c)	(d)
26.	(a)	(b)	(c)	(d)		62.	(a)	(b)	(c)	(d)		98.	(a)	(b)	(c)	(d)
27.	(a)	(b)	(c)	(d)		63.	(a)	(b)	(c)	(d)		99.	(a)	(b)	(c)	(d)
28.	(a)	(b)	(c)	(d)		64.	(a)	(b)	(c)	(d)		100.	(a)	(b)	(c)	(d)
29.	(a)	(b)	(c)	(d)		65.	(a)	(b)	(c)	(d)						
30.	(a)	(b)	(c)	(d)		66.	(a)	(b)	(c)	(d)						
31.	(a)	(b)	(c)	(d)		67.	(a)	(b)	(c)	(d)						
32.	(a)	(b)	(c)	(d)		68.	(a)	(b)	(c)	(d)						
33.	(a)	(b)	(c)	(d)		69.	(a)	(b)	(c)	(d)						
34.	(a)	(b)	(c)	(d)		70.	(a)	(b)	(c)	(d)						
35.	(a)	(b)	(c)	(d)		71.	(a)	(b)	(c)	(d)						
36.	(a)	(b)	(c)	(d)		72.	(a)	(b)	(c)	(d)						

Radiography Practice Test I

101.	ⓐ	ⓑ	ⓒ	ⓓ	137.	ⓐ	ⓑ	ⓒ	ⓓ	173.	ⓐ	ⓑ	ⓒ	ⓓ
102.	ⓐ	ⓑ	ⓒ	ⓓ	138.	ⓐ	ⓑ	ⓒ	ⓓ	174.	ⓐ	ⓑ	ⓒ	ⓓ
103.	ⓐ	ⓑ	ⓒ	ⓓ	139.	ⓐ	ⓑ	ⓒ	ⓓ	175.	ⓐ	ⓑ	ⓒ	ⓓ
104.	ⓐ	ⓑ	ⓒ	ⓓ	140.	ⓐ	ⓑ	ⓒ	ⓓ	176.	ⓐ	ⓑ	ⓒ	ⓓ
105.	ⓐ	ⓑ	ⓒ	ⓓ	141.	ⓐ	ⓑ	ⓒ	ⓓ	177.	ⓐ	ⓑ	ⓒ	ⓓ
106.	ⓐ	ⓑ	ⓒ	ⓓ	142.	ⓐ	ⓑ	ⓒ	ⓓ	178.	ⓐ	ⓑ	ⓒ	ⓓ
107.	ⓐ	ⓑ	ⓒ	ⓓ	143.	ⓐ	ⓑ	ⓒ	ⓓ	179.	ⓐ	ⓑ	ⓒ	ⓓ
108.	ⓐ	ⓑ	ⓒ	ⓓ	144.	ⓐ	ⓑ	ⓒ	ⓓ	180.	ⓐ	ⓑ	ⓒ	ⓓ
109.	ⓐ	ⓑ	ⓒ	ⓓ	145.	ⓐ	ⓑ	ⓒ	ⓓ	181.	ⓐ	ⓑ	ⓒ	ⓓ
110.	ⓐ	ⓑ	ⓒ	ⓓ	146.	ⓐ	ⓑ	ⓒ	ⓓ	182.	ⓐ	ⓑ	ⓒ	ⓓ
111.	ⓐ	ⓑ	ⓒ	ⓓ	147.	ⓐ	ⓑ	ⓒ	ⓓ	183.	ⓐ	ⓑ	ⓒ	ⓓ
112.	ⓐ	ⓑ	ⓒ	ⓓ	148.	ⓐ	ⓑ	ⓒ	ⓓ	184.	ⓐ	ⓑ	ⓒ	ⓓ
113.	ⓐ	ⓑ	ⓒ	ⓓ	149.	ⓐ	ⓑ	ⓒ	ⓓ	185.	ⓐ	ⓑ	ⓒ	ⓓ
114.	ⓐ	ⓑ	ⓒ	ⓓ	150.	ⓐ	ⓑ	ⓒ	ⓓ	186.	ⓐ	ⓑ	ⓒ	ⓓ
115.	ⓐ	ⓑ	ⓒ	ⓓ	151.	ⓐ	ⓑ	ⓒ	ⓓ	187.	ⓐ	ⓑ	ⓒ	ⓓ
116.	ⓐ	ⓑ	ⓒ	ⓓ	152.	ⓐ	ⓑ	ⓒ	ⓓ	188.	ⓐ	ⓑ	ⓒ	ⓓ
117.	ⓐ	ⓑ	ⓒ	ⓓ	153.	ⓐ	ⓑ	ⓒ	ⓓ	189.	ⓐ	ⓑ	ⓒ	ⓓ
118.	ⓐ	ⓑ	ⓒ	ⓓ	154.	ⓐ	ⓑ	ⓒ	ⓓ	190.	ⓐ	ⓑ	ⓒ	ⓓ
119.	ⓐ	ⓑ	ⓒ	ⓓ	155.	ⓐ	ⓑ	ⓒ	ⓓ	191.	ⓐ	ⓑ	ⓒ	ⓓ
120.	ⓐ	ⓑ	ⓒ	ⓓ	156.	ⓐ	ⓑ	ⓒ	ⓓ	192.	ⓐ	ⓑ	ⓒ	ⓓ
121.	ⓐ	ⓑ	ⓒ	ⓓ	157.	ⓐ	ⓑ	ⓒ	ⓓ	193.	ⓐ	ⓑ	ⓒ	ⓓ
122.	ⓐ	ⓑ	ⓒ	ⓓ	158.	ⓐ	ⓑ	ⓒ	ⓓ	194.	ⓐ	ⓑ	ⓒ	ⓓ
123.	ⓐ	ⓑ	ⓒ	ⓓ	159.	ⓐ	ⓑ	ⓒ	ⓓ	195.	ⓐ	ⓑ	ⓒ	ⓓ
124.	ⓐ	ⓑ	ⓒ	ⓓ	160.	ⓐ	ⓑ	ⓒ	ⓓ	196.	ⓐ	ⓑ	ⓒ	ⓓ
125.	ⓐ	ⓑ	ⓒ	ⓓ	161.	ⓐ	ⓑ	ⓒ	ⓓ	197.	ⓐ	ⓑ	ⓒ	ⓓ
126.	ⓐ	ⓑ	ⓒ	ⓓ	162.	ⓐ	ⓑ	ⓒ	ⓓ	198.	ⓐ	ⓑ	ⓒ	ⓓ
127.	ⓐ	ⓑ	ⓒ	ⓓ	163.	ⓐ	ⓑ	ⓒ	ⓓ	199.	ⓐ	ⓑ	ⓒ	ⓓ
128.	ⓐ	ⓑ	ⓒ	ⓓ	164.	ⓐ	ⓑ	ⓒ	ⓓ	200.	ⓐ	ⓑ	ⓒ	ⓓ
129.	ⓐ	ⓑ	ⓒ	ⓓ	165.	ⓐ	ⓑ	ⓒ	ⓓ					
130.	ⓐ	ⓑ	ⓒ	ⓓ	166.	ⓐ	ⓑ	ⓒ	ⓓ					
131.	ⓐ	ⓑ	ⓒ	ⓓ	167.	ⓐ	ⓑ	ⓒ	ⓓ					
132.	ⓐ	ⓑ	ⓒ	ⓓ	168.	ⓐ	ⓑ	ⓒ	ⓓ					
133.	ⓐ	ⓑ	ⓒ	ⓓ	169.	ⓐ	ⓑ	ⓒ	ⓓ					
134.	ⓐ	ⓑ	ⓒ	ⓓ	170.	ⓐ	ⓑ	ⓒ	ⓓ					
135.	ⓐ	ⓑ	ⓒ	ⓓ	171.	ⓐ	ⓑ	ⓒ	ⓓ					
136.	ⓐ	ⓑ	ⓒ	ⓓ	172.	ⓐ	ⓑ	ⓒ	ⓓ					

Patient Care and Education

Choose the best answer from the choices given.

1. If a patient is placed on airborne isolation, which of the following protective garment(s) must be worn prior to entering the room?

    1. mask
    2. gloves
    3. booties
    4. gown

    a. 1 and 2 only
    b. 1 and 4 only
    c. 1, 2, and 4 only
    d. all of the above

2. When performing a two-person resuscitation on a 12-year-old patient, the ratio of compressions to breaths are
    a. 30 compressions to two breaths.
    b. 30 compressions to three breaths.
    c. 15 compressions to two breaths.
    d. 20 compressions to two breaths.

3. Prior to a small bowel series, the patient must be NPO for how many hours?
    a. 6–12
    b. 8–10
    c. 8–12
    d. 6–8

4. A pulse greater than 100 beats per minute is considered
    a. brachycardic.
    b. tachycardic.
    c. hypoglycemic.
    d. hypertensive.

5. Which of the following is true when referring to the use of positive contrast agents?
    a. It decreases the attenuation and decreases the density.
    b. It increases the attenuation and decreases the density.
    c. It increases the attenuation and increases the density.
    d. It decreases the attenuation and increases the density.

6. What does the term *diastolic* mean?
    a. measurement of pressure of the heart during contraction
    b. fever
    c. heart failure
    d. measurement of pressure of the heart during relaxation

7. What does the term *dyspnea* mean?
    a. difficulty sleeping
    b. difficulty eating
    c. difficulty breathing
    d. difficulty seeing

8. If a patient presents with low blood sugar and a rapid onset, he or she is considered to be
    a. hyperglycemic.
    b. hyposystemic.
    c. hypoglycemic.
    d. tachycardic.

9. Which route of transmission would best describe a disease or infection that is transmitted via an animal?
    a. airborne
    b. vector
    c. droplet
    d. indirect

10. An advanced directive should be filled out by the patient before he or she
  a. becomes incapacitated.
  b. is discharged from the hospital.
  c. returns to work.
  d. eats lunch.

11. Which of the following items must be present on a radiograph?

  1. name
  2. date
  3. left or right markers
  4. medical record number

  a. 1 and 2 only
  b. 1, 2, and 3 only
  c. 1, 2, and 4 only
  d. all of the above

12. The minimum height of an IV bag should be no less than
  a. 24 inches.
  b. 18 inches.
  c. 12 inches.
  d. 10 inches.

13. At what temperature should the water be when mixed with powder for a barium enema?
  a. 98.6°F
  b. 99°F
  c. 100°F
  d. 100.4°F

14. What does the term *systolic* mean?
  a. measurement of pressure of the heart during relaxation
  b. measurement of pressure of oxygen per minute
  c. measurement of pressure of the lungs during respiration
  d. measurement of pressure of the heart during contraction

15. When performing a surgical cholangiogram, which of the following routes of administration should be utilized?
  a. direct injection
  b. intravenous injection
  c. intramuscular injection
  d. subcutaneous injection

16. In the event that a patient experiences a severe loss of blood, there is a great possibility that he or she may go into which type of shock?
  a. hyperventilation
  b. hypovolemic
  c. hypodermic
  d. hypergenic

17. Air is considered what type of contrast agent?
  a. positive
  b. negative
  c. iodinated ionic
  d. iodinated nonionic

18. A delayed reaction to nonionic iodinated contrast can occur up until
  a. five minutes postinjection.
  b. 1 hour postinjection.
  c. 8 hours postinjection.
  d. 24 hours postinjection.

19. Surgical asepsis is the
  a. spreading of organisms during surgery.
  b. complete removal of organisms from equipment and environment.
  c. surgical removal of microorganisms.
  d. surgical removal of organisms from patient.

20. Which of the following is true regarding the use of negative contrast agents?
   a. They decrease the attenuation and decrease the density.
   b. They increase the attenuation and decrease the density.
   c. They increase the attenuation and increase the density.
   d. They decrease the attenuation and increase the density.

21. When the possibility of a perforated ulcer or ruptured appendix exists, which type of contrast should be used?
   a. ionic contrast
   b. nonionic contrast
   c. aqueous iodine compounds
   d. barium sulfate

22. Which type of contrast media is associated with a higher risk of reactions?
   a. ionic
   b. nonionic
   c. gadolinium
   d. aqueous iodine

23. Nausea, vomiting, hives, and flushed skin are all symptoms of what type of reaction?
   a. anaphylactic
   b. cardiogenic
   c. neurogenic
   d. hyperglycemic

24. A mechanical breathing device that assists patients when they cannot breathe on their own is called
   a. a tracheostomy.
   b. a re-breather mask.
   c. an oxygen tank.
   d. a ventilator.

25. Which definition best describes a compound fracture?
   a. bone is protruding through the skin
   b. the fracture of a bone in two places
   c. one fragment is firmly driven into the other
   d. fracture of the distal radius with posterior displacement

26. In what order should the following exams be scheduled?
   1. upper GI series
   2. intravenous pyelogram
   3. barium enemwa
   4. femur x-ray

   a. 4, 3, 1, 2
   b. 1, 2, 3, 4
   c. 4, 2, 3, 1
   d. 3, 2, 4, 1

Equipment Operation and Maintenance

Choose the best answer from the choices given.

27. What is true of a capacitor discharge portable machine?
   a. It does not have to be plugged in.
   b. It stores enough energy for about 50 chest exams.
   c. It must be charged immediately before each exposure.
   d. It allows higher mA and kV than other types.

28. What is NOT another name for coherent scatter?
   a. adherent scatter
   b. Rayleigh scatter
   c. classical scatter
   d. All of the above mean coherent scatter.

29. A rotating anode works off of _____ motor.
a. an indirect
b. a direct
c. an induction
d. a noninduction

30. Which interaction does not occur at the diagnostic range?
a. pair production
b. coherent
c. photoelectric
d. Compton

31. The focusing cup keeps the electrons
a. together.
b. apart.
c. positively charged.
d. none of the above.

32. For phototiming, backup time should be set at ____% of the anticipated manual exposure.
a. 50
b. 100
c. 150
d. 200

33. ABC stands for
a. anode blooming control.
b. automatic brightness control.
c. automatic brightness center.
d. none of the above.

34. Which of the following comes first in the image intensifier?
a. input phosphor
b. photocathode
c. electrostatic lenses
d. output phosphor

35. Which of the following occurs during characteristic radiation?
a. heat
b. a projectile electron
c. a negatron
d. a positron

36. The cathode has a ____ charge, and the anode has a ____ charge.
a. positive/negative
b. neutral/negative
c. negative/positive
d. none of the above

37. AEC stands for
a. anti-exposure control.
b. anode-electron compressor.
c. automatic exposure control.
d. none of the above.

38. The term *capacitor discharge* is most associated with
a. fluoroscopy.
b. mobile radiography.
c. tube current.
d. none of the above.

39. All of the following are kVp voltage accuracy tests EXCEPT
a. Wisconsin test.
b. digital kilovolt meter.
c. kilo-equivalent method.
d. Ardan and Crooke.

40. The most popular type of exposure activation switch for fluoro use is the
a. relay switch.
b. circuit switch.
c. microswitch.
d. dead man switch.

41. What does the output phosphor of an image intensifier emit?
    a. electrons
    b. light
    c. x-rays
    d. none of the above

42. In German, *bremsstrahlung* means
    a. heated.
    b. braking.
    c. electric.
    d. speeding.

43. The gas found in ion chambers of AEC devices is
    a. nitrogen.
    b. oxygen.
    c. Freon.
    d. free air.

44. What alloy is commonly used in combination with tungsten in the construction of an anode?
    a. copper
    b. tin
    c. aluminum
    d. rhenium

45. Within the image intensifier, the purpose of the photocathode is to
    a. to change electrons to x-rays.
    b. to change light to electrons.
    c. to change electrons to light.
    d. none of the above.

46. The device that allows free flow of electrons is a(n)
    a. insulator.
    b. transducer.
    c. conductor.
    d. resistor.

47. Why is there a warm-up procedure for x-ray machines?
    a. to increase the production of the tube
    b. to strengthen the space charge
    c. to properly warm the anode
    d. all of the above

48. What interacts with the input phosphor of the image intensification tube?
    a. scatter photons
    b. remnant photons
    c. absorbed photons
    d. none of the above

49. All of the following are parts of the television camera tube *except* the
    a. target plate.
    b. electron gun.
    c. steering coils.
    d. photocathode.

50. Which of the following is NOT associated with the anode?
    a. target
    b. stem
    c. focusing cup
    d. stator

Image Production and Evaluation

Choose the best answer from the choices given.

51. Which of the following choices would effectively reduce the amount of magnification of an object being imaged?
   a. having a greater OID present
   b. utilizing a parallel central ray
   c. having the least amount of OID possible
   d. decreasing the SID

52. Which formula could a technologist utilize in order to maintain density on an image if the source to image distance must be adjusted?
   a. the inverse square law
   b. MF = SID/SOD
   c. the direct square law
   d. P = FSS × (OID/SOD)

53. Of the following sets of factors, which would produce the greatest amount of focal spot blur?
   a. 46" SID, 3" OID, 0.6 FSS
   b. 40" SID, 5" OID, 1.2 FSS
   c. 50" SID, 2" OID, 0.3 FSS
   d. 54" SID, 8" OID, 1.2 FSS

54. If it becomes necessary to increase the density of a film by a factor of 2, which of the following choices would allow a radiographer to do so?
   a. increase kVp by 15%
   b. increase kVp by 5
   c. increase kVp by 30%
   d. increase grid ratio to 8:1

55. By increasing the energy of an x-ray beam, the resulting radiograph will exhibit
   a. decreased density.
   b. decreased contrast.
   c. increased focal spot blur.
   d. increased contrast.

56. The primary benefit in using a rare earth intensifying screen is that it
   a. decreases image noise.
   b. allows for greater densities.
   c. allows for decreases in patient exposure.
   d. has no benefit.

57. The majority of scattered radiation is produced by
   a. the cassette cover.
   b. the tube.
   c. the table.
   d. the patient.

58. What mA station would need to be selected if 40 mAs were needed for a given exposure lasting 0.08 seconds?
   a. 300 mA
   b. 400 mA
   c. 500 mA
   d. 600 mA

59. In which of the following technical factors would the greatest amount of magnification be present?
   a. 40" SID, 2" OID
   b. 44" SID, 4" OID
   c. 60" SID, 8" OID
   d. 72" SID, 5" OID

60. As the mAs for a given examination decreases, what is the effect on density?
a. decreases proportionally
b. increases proportionally
c. decreases by half when reduced by 15%
d. decreases by half when reduced by 30%

61. An examination is performed using 32 mAs and 75 kVp with a 16:1 grid. If the exam was to be repeated utilizing a 6:1 grid, what would be the new mAs?
a. 8 mAs
b. 16 mAs
c. 20 mAs
d. 32 mAs

62. Rare earth screen/film combinations are utilized by a facility. What is the type of safelight best associated with this combination?
a. Wratten 6B
b. Hurter 3B
c. Kodak GBX
d. Driffeld 6B

63. In an automatic processor, the action of the developer is neutralized by the
a. wash.
b. dryer.
c. fixer.
d. guide shoes.

64. Which of the following factors would produce the greatest recorded detail?
a. using high kilovoltage
b. using longer time settings
c. using a small focal spot
d. using a high ratio grid

65. Which of the following tissue types has the greatest differential absorption?
a. fat
b. water
c. soft tissue
d. bone

66. The thickness of absorbing material necessary in order to cut the intensity of the beam to half of its original value refers to
a. half value layer.
b. filtration.
c. air gap.
d. a grid.

67. The base in radiographic film is commonly made up of what type of material?
a. aluminum
b. molybdenum
c. polyester
d. pyrex

68. Despite the great amount of detail attained when utilizing direct exposure film, what is its primary disadvantage?
a. cost
b. increased dosage to patient
c. difficult to use
d. space consuming

69. Which of the following is an advantage when using high ratio grids?
a. degree of scatter cleanup
b. degree of primary beam absorbed
c. lack of positioning latitude
d. exposure required

70. Which of the following are considered rare earth phosphors?

    1. gadolinium
    2. yttrium
    3. lanthanum

a. 1 only
b. 1 and 3 only
c. 1 and 2 only
d. 1, 2, and 3

71. Which of the following factors are necessary in order for quantum mottle to be present on an image?

    1. fast speed screen
    2. high kVp technique
    3. low mAs technique

a. 1 only
b. 1 and 2 only
c. 1 and 3 only
d. 1, 2, and 3

72. The overall blackening present on a radiograph is the definition of
a. density.
b. contrast.
c. spatial resolution.
d. moiré effect.

73. Which of the following factors can be used in order to regulate radiographic density?

    1. exposure time
    2. milliamperage
    3. kilovoltage

a. 2 only
b. 1 and 2 only
c. 2 and 3 only
d. 1, 2, and 3

74. How much exposure time would be necessary if 600 mA were utilized in order to produce 32 mA?
a. 0.025 sec
b. 0.05 sec
c. 0.10 sec
d. 0.50 sec

75. What would the mA be if 0.10 seconds were used in order to produce 60 mA?
a. 200 mA
b. 400 mA
c. 600 mA
d. 000 mA

76. The term *short-scale contrast* generally refers to a film which contains
a. few densities with great differences between them.
b. a large amount of densities with less differences between them.
c. the same density present throughout the entire image.
d. no density present throughout the entire image.

77. If the kilovoltage had to be increased in order to assist in the reduction of patient dosage, the most noticeable effect on the finished image would be
a. decreased contrast.
b. increased contrast.
c. increased light fog.
d. decreased density.

78. As the speed of film increases, if utilizing the same technique as a slow speed film, what would happen to the density on the film?
a. It would increase.
b. It would decrease.
c. It would remain the same.
d. Speed has no relationship to density.

79. In an exposure, filtration serves what main purpose?
a. It reduces technologist dosage.
b. It increases patient dosage.
c. It decreases patient dosage.
d. It reduces scattered radiation production.

80. A film was produced utilizing 75 kVp and 5 mAs. In order to decrease dosage to the patient, the film could be repeated utilizing which technique in order to maintain density?
a. 90 kVp, 2.5 mAs
b. 90 kVp, 10 mAs
c. 65 kVp, 10 mAs
d. 90 kVp, 5 mAs

81. When performing a study, if the SID must be increased by double the original distance, what effect will this have on density?
a. Density would double.
b. Density would be halved.
c. Density would be reduced by a factor of 4.
d. Density would be the same.

82. What is the magnification factor if an object 6 inches in height is imaged from 44 inches?
a. 0.6
b. 1
c. 1.2
d. 1.8

83. If an exam is performed at 5 mAs utilizing an 800 speed screen, what would the new mAs be if the exam was performed utilizing a 100 speed screen?
a. 5 mAs
b. 10 mAs
c. 20 mAs
d. 40 mAs

84. Utilizing a type of beam restrictor will have which of the following effects on radiographic contrast?
a. It increases.
b. It decreases by a factor of 2.
c. It has no effect.
d. It decreases proportionally.

85. Which is an example of an additive pathology that requires an increase in the technique used?
a. osteomalacia
b. pneumonia
c. emphysema
d. multiple myeloma

86. Film should be stored in a location that has which of the following humidity levels?
a. 0–20%
b. 40–60%
c. 80–100%
d. does not matter

87. The term *quantity*, when pertaining to the x-ray beam, commonly refers to which technical factor?
a. mAs
b. kVp
c. grid ratio
d. object to image distance

88. Which of the following effects would an image have if there was a decrease in the photon energy in an x-ray beam?
a. increased contrast
b. decreased contrast
c. decreased density
d. no effect

89. In order to produce a change equaling double the original density on an image, the kVp must be manipulated at least
   a. 5%.
   b. 15%.
   c. 20%.
   d. 30%.

90. The photostimulable phosphor relays its information after being exposed by a laser beam into which type of device?
   a. digital to analog converter
   b. electrostatic lens
   c. cathode ray tube
   d. analog to digital converter

91. What would the mA be if 0.25 seconds was used in order to produce 100 mA?
   a. 200 mA
   b. 400 mA
   c. 600 mA
   d. 1,000 mA

92. What is the mA value if 800 mA was used at 0.05 seconds?
   a. 10 mA
   b. 20 mA
   c. 30 mA
   d. 40 mA

93. Which of the groups of exposure factors will produce the greatest radiographic density?
   a. 100 mA, 0.30 sec
   b. 200 mA, 0.10 sec
   c. 400 mA, 0.03 sec
   d. 600 mA, 0.03 sec

94. Exposure factors of 90 kVp and 2.5 mAs are used for a particular nongrid exposure. What should be the new mAs if a 5:1 grid is added?
   a. 5 mAs
   b. 10 mAs
   c. 2.5 mAs
   d. 1.25 mAs

95. Which of the following groups of exposure factors would be most appropriate to control involuntary motion?
   a. 400 mA, 0.02 sec
   b. 500 mA, 0.1 sec
   c. 200 mA, 0.01 sec
   d. 600 mA, 0.33 sec

96. For compensation, when using the 15% rule, if the kilovolts are increased by 15%, the mAs should be
   a. doubled.
   b. reduced by one-half.
   c. reduced by 15%.
   d. quadrupled.

97. Which of the following is a factor that controls shape distortion?
   a. milliamperage-seconds
   b. beam-part-film alignment
   c. SID
   d. FSS

98. Which of the following will cause a loss of recorded detail?
   a. mAs changes
   b. kVp changes
   c. high grid ratios
   d. patient motion

99. Which of the following would affect shape distortion?
   a. SID
   b. OID
   c. SOD
   d. tube angulation

100. The most common filtering material for an x-ray beam in the diagnostic range is
   a. thorium.
   b. aluminum.
   c. copper.
   d. lead.

Radiographic Procedures

Choose the best answer from the choices given.

101. In order to rule out a fracture of the base of the fifth metatarsal, the best projection that could be utilized is the
   a. medial oblique projection.
   b. plantodorsal projection.
   c. dorsoplantar projection.
   d. lateral projection.

102. How many vertebrae constitute the entire spinal column?
   a. 23
   b. 28
   c. 33
   d. 37

103. Which carpal bones are best demonstrated when performing a posteroanterior oblique projection of the right wrist?
   a. pisiform and hamate
   b. greater multangular and lesser multangular
   c. scaphoid and trapezoid
   d. capitate and lunate

104. The right sacroiliac joint would be best demonstrated in which projection?
   a. Chassard-Lapine projection
   b. AP projection
   c. RAO projection
   d. RPO projection

105. Which position is best in order to demonstrate the zygapophyseal articulations of the lumbar spine?
   a. AP projection
   b. AP 45 degree oblique projection
   c. lateral projection
   d. AP erect projection

106. Which plane, used frequently in positioning, separates the body into equal anterior and posterior portions?
   a. midcoronal plane
   b. midsagittal plane
   c. transverse plane
   d. oblique plane

107. The 35–45 degree posterior oblique projection is used to best demonstrate which specific area of the shoulder girdle free of superimposition?
   a. scapula
   b. AC joint
   c. glenoid fossa
   d. humeral head

108. In order to best visualize the intervertebral space between C-7/T-1, especially in patients who cannot fully depress their shoulders, the position best used is the
   a. AP.
   b. swimmer's.
   c. anterior oblique.
   d. posterior oblique.

109. How many posterior ribs should be visualized on a properly performed PA chest radiograph?
  a. 6
  b. 8
  c. 10
  d. 12

110. How much central ray angulation is required if the sacrum is the desired area of interest?
  a. 15 degrees caudal
  b. 15 degrees cephalic
  c. 25 degrees cephalic
  d. 25 degrees caudal

111. What is the largest carpal bone?
  a. navicular
  b. pisiform
  c. trapezoid
  d. capitate

112. In order to best minimize the magnification of the structures in the mediastinum, what distance should be obtained when performing a posteroanterior chest radiograph?
  a. 36 inches
  b. 40 inches
  c. 48 inches
  d. 72 inches

113. In positioning the abdominal serried, which of the following projections could be utilized to best demonstrate possible free air?

  1. KUB
  2. left lateral decubitis
  3. upright

  a. 1 only
  b. 1 and 2 only
  c. 2 and 3 only
  d. 1, 2, and 3

114. Fractures attributed to the shoulder commonly take place at which aspect of the humerus?
  a. surgical neck
  b. anatomical neck
  c. humeral head
  d. glenoid fossa

115. When attempting to visualize the occipital bone through a patient's open mouth, which two anatomical structures need to be adjusted to lie in the same vertical plane?
  a. vomer and ethmoid sinus
  b. mastoid tip and lower edge of the upper incisors
  c. EAM and the lower edge of the upper incisors
  d. EAM and the hard palate

116. When there is a suspected fracture of the right axillary ribs, the best projection to be performed is
  a. LAO.
  b. LPO.
  c. PA.
  d. AP.

117. The articulations between the zygapophyseal joints of the cervical vertebrae are best seen in which projection?
  a. anteroposterior
  b. lateral
  c. open mouth
  d. PA obliques

118. Which carpal bone is considered the most frequently fractured, especially among roller bladers?
  a. scaphoid
  b. triquetrum
  c. hamate
  d. capitate

119. If a patient arrives with suspected bilateral hip fractures, after the AP projection is performed, which method could a technologist use in order to best attempt a lateral projection?
a. Cleaves Method
b. Modified Cleaves Method
c. Clements-Nakayama Method
d. Waters Method

120. The patellofemoral joint is best visualized free of the femur in which of the following methods?
a. Grashey
b. Settegast
c. Judet
d. Lawrence

121. The plane of the body which separates the body into equal right and left halves is the
a. midcoronal plane.
b. midsagittal plane.
c. transverse plane.
d. oblique plane.

122. In cases where visualizing the dens through a patient's open mouth proves too difficult due to the patient's condition, the dens could be visualized through which structure if the Fuchs Method was utilized?
a. foramen magnum
b. sphenoid sinus
c. obturator foramen
d. open mouth

123. An image is presented showing the stomach with barium within the fundus while the body and pylorus are filled with air. The position demonstrated is the
a. AP oblique LPO.
b. PA.
c. PA oblique RAO.
d. PA Trendelenburg.

124. What is the bony protuberance that can be palpated on the anterior tibia?
a. tibial spine
b. tibial eminence
c. tibial plateau
d. tibial tuberosity

125. Which of the following anatomical structures are located in the left upper quadrant of the abdomen?
1. Stomach
2. Liver
3. Splenic flexure

a. 1 only
b. 2 only
c. 1 and 3 only
d. 1, 2, and 3

126. Which of the following projections best demonstrates the intercondylar fossa of the knee?
1. Holmblad
2. Camp Coventry
3. Pearson

a. 1 only
b. 2 only
c. 1 and 2 only
d. 1, 2, and 3

127. With the patient in the Sims position, how far should the catheter tip be inserted upon the start of a barium enema study?
a. 2–3 inches
b. 6 inches
c. 1 inch
d. 3.5–4 inches

128. When an image of the shoulder presents the lesser tubercle in profile medially, what position was the patient in?
  a. AP neutral
  b. AP internal rotation
  c. AP external rotation
  d. AP oblique

129. The flexure where the transverse colon meets the descending colon is called the
  a. hepatic flexure.
  b. hiatus flexure.
  c. gastric flexure.
  d. splenic flexure.

130. Which of the positions for paranasal sinuses would best display all four of the sinus groups?
  a. Parietoacanthial projection (Waters)
  b. AP axial projection (Caldwell)
  c. submentovertex
  d. lateral

131. Within the chest, what is the term for the portion of the lungs that is projected superior to the clavicles?
  a. cardiophrenic angle
  b. carina
  c. costophrenic angle
  d. apex

132. How many bones are in the normal human body?
  a. 185
  b. 200
  c. 206
  d. 212

133. The common way of localizing the acetabulum of the hip joint is to bisect an imaginary line between what two structures?
  a. symphysis pubis and iliac crest
  b. iliac crest and ASIS
  c. symphysis pubis and ASIS
  d. iliac crest and pelvic inlet

134. When projecting the orbits using the Rheese Method, the proper position has been obtained when the optic foramen is projected into which quadrant?
  a. upper lateral quadrant
  b. upper medial quadrant
  c. lower lateral quadrant
  d. lower medial quadrant

135. It is absolutely necessary that when positioning a patient for sinus studies that all four projections are taken with the patient
  a. upright.
  b. supine.
  c. Trendelenberg.
  d. prone.

136. Nutrients are most absorbed into the body by what part of the digestive system?
  a. esophagus
  b. small intestine
  c. stomach
  d. rectum

137. Urine that is produced in the kidney is transported to the bladder via the
  a. ureter.
  b. urethra.
  c. aorta.
  d. inferior vena cava.

138. In the typical lumbar spine, how many vertebrae are present?
a. 5
b. 7
c. 12
d. 8

139. The external portion of the ear is commonly called the
a. auricle.
b. EAM.
c. petrous portion.
d. mastoid process.

140. When performing a barium enema, if the splenic flexure of the colon is of interest, then the projection that will best demonstrate this structure is the
a. PA.
b. LPO.
c. RPO.
d. AP axial.

141. Which of the following is not a cranial bone?
a. vomer
b. occipital
c. parietal
d. temporal

142. The dens is related to which vertebrae?
a. C-1
b. C-2
c. C-7
d. L-1

143. For visualization of the occipital bone without the superimposition of the other cranial structures, the midsagittal plane of the head and the baseline IOML should be placed perpendicular to the IR. How should the central ray be directed?
a. 20 degrees caudad
b. 30 degrees caudad
c. 37 degrees caudad
d. 45 degrees caudad

144. Which part of the mandible articulates with the temporal bone of the cranium?
a. angle
b. body
c. condylar process
d. coronoid process

145. What type of curvature does the thoracic spine present?
a. lordotic
b. kyphotic
c. linear
d. transverse

146. When the navicular bone and its articulations are of interest when radiographing the wrist, which of the following positions is best?
a. lateral
b. Stecher
c. carpal canal
d. AP

147. The sthenic body habitus makes up what percentage of the population?
a. 5%
b. 35%
c. 10%
d. 50%

148. At C-2, the vertical process that extends upward and articulates with C-1 is also referred to as the
a. styloid process.
b. coronoid.
c. odontoid.
d. superior articulating process.

149. The motion provided by peristalsis is strongest at which aspect of the digestive system?
a. esophagus
b. stomach
c. small intestine
d. colon

150. The baseline, which is an imaginary line drawn from the external auditory meatus (EAM) to the outer canthus of the eye is called
a. interpupillary line.
b. orbitomeatal line.
c. ancanthiomeatal line.
d. mentomeatal line.

151. In a projection of the shoulder in which the patient is in an AP position with the body rotated 45 degrees toward the shoulder of interest, the part being demonstrated is the
a. glenoid fossa.
b. body of the scapula.
c. AC joint.
d. greater tuberosity.

152. In the lateral skull projection, the baseline that is perpendicular to the plane of the IR is the
a. OML.
b. IPL.
c. IOML.
d. GML.

153. The femur is classified as a
a. sesamoid.
b. long bone.
c. irregular bone.
d. flat bone.

154. The projection that best demonstrates the occipital bone of the cranium is the
a. AP projection.
b. lateral projection.
c. AP axial Townes Projection.
d. submento vertical projection.

155. The projection of the paranasal sinuses which best demonstrates the maxillary sinus is the
a. lateral projection.
b. AP axial projection (Caldwell Method).
c. parietoacanthial projection (Waters Method).
d. submentovertical projection.

156. The three aspects of the small intestine in order starting from the stomach are
a. ilium, jejunum, duodenum.
b. jejunum, ilium, duodenum.
c. duodenum, ilium, jejunum.
d. duodenum, jejunum, ilium.

157. In order to best demonstrate the lower ribs, the exposure should be made
a. when the patient takes a full deep breath and holds it.
b. on the second full inspiration.
c. on full exhalation.
d. while the patient is utilizing a breathing technique.

158. If the patient is placed in an LAO position with the pelvis rotated 25 degrees and the center ray exiting 1" medial to the elevated ASIS, what is the resulting structure?
a. right sacroiliac joint
b. left sacroiliac joint
c. left ischial rami
d. right ischial rami

159. For a parietoacanthial projection (Waters Method) of the sinuses, the orbitomeatal line should form a _____ angle to the plane of the IR.
a. 25 degree
b. 30 degree
c. 37 degree
d. 40 degree

160. During a barium enema study, in order to facilitate the proper flow of contrast into the colon, the enema bag should be located
a. 18"–24" above the anus.
b. 10"–17" above the anus.
c. 3.5"–4" above the anus.
d. 12"–18" below the anus.

Radiation Protection

Choose the best answer from the choices given.

161. Factors which determine protective barrier thicknesses include all of the following EXCEPT
a. use.
b. workload.
c. time occupancy.
d. intensity.

162. When operating a mobile x-ray unit, which of the following is NOT an acceptable source to skin distance?
a. 10 inches
b. 12 inches
c. 8 inches
d. both a and c

163. Which of the following are grounds for a repeat radiograph?
a. underexposure of area of interest
b. poor positioning of patient
c. mislabeling of the film (patient identification)
d. all of the above

164. Leakage radiation on a mobile x-ray unit is a concern for
a. the patient.
b. the doctor.
c. the radiologic technologist.
d. none of the above.

165. Which of the following devices is NOT located between the patient and the x-ray tube?
a. primary beam
b. filtration
c. grid
d. collimator

166. Which of the following is NOT intended to touch or be placed on the patient?
a. flat contact shield
b. fig leaf shield
c. shaped contact shield
d. shadow shield

167. Scatter radiation is created any time
    a. a grid is used.
    b. you use too much kVp.
    c. x-rays pass through matter.
    d. none of the above

168. The GSD for the general population cannot exceed_____ rem(s).
    a. 3
    b. 1
    c. 2
    d. 5

169. The GSD for the occupational population cannot exceed_____ rem(s).
    a. 3
    b. 1
    c. 2
    d. 5

170. What is your best protection from scatter when performing a portable radiograph?
    a. to stand as far away as possible
    b. to collimate to the size of the film
    c. to set a higher KV if possible
    d. to use high-speed rotation if available

171. Redness of the skin from radiation exposure is referred to as
    a. exfoliation.
    b. purulent.
    c. erythema.
    d. erythrocyte.

172. The occupation dose (of radiation workers) is expressed using what unit of measurement?
    a. rem
    b. RAD
    c. LET
    d. roentgen

173. What is NOT a long-term radiation effect?
    a. leukemia
    b. lowered WBC count
    c. cataracts
    d. life-span shortening

174. Which of the following is the best way to reduce exposure to the reproductive organs of a man?
    a. beam collimation
    b. proper filtration
    c. fast screen speeds
    d. high kilovoltage

175. Which type of blood cell is known as the least radioresistant?
    a. lymphocyte
    b. erythrocyte
    c. thrombocyte
    d. platelet

176. Why do film badges have filters?
    a. to prevent overheating during exposure
    b. to increase the radiation during exposure
    c. to indicate the type of radiation received
    d. to shield the badge from radiation

177. Which type of dosimeter is heated so it can be read?
    a. TLD
    b. pocket dosimeter
    c. film badge
    d. ionization chamber

178. Which of the following is NOT a high-risk examination for a pregnant woman?
    a. IVP
    b. shoulder
    c. pelvis/hip
    d. UGI series

179. Which technique will produce the lowest patient dose on a spine radiograph?
  a. single phase, 80 kVp
  b. single phase, 65 kVp
  c. three phase, 65 kVp
  d. three phase, 80 kVp

180. Sievert is a known radiation measurement of
  a. exposure.
  b. dose equivalence.
  c. activity.
  d. dose.

181. Which scatter reduction device will require the lowest dose to a patient's skull?
  a. air gap
  b. 8:1 grid
  c. 6:1 grid
  d. 16:1 grid

182. The effects of radiation are known to be influenced by all of the following EXCEPT
  a. dose rate.
  b. size of the cell(s).
  c. type of cell.
  d. type of radiation.

183. The effects of radiation are usually more profound if the exposure is delivered
  a. in one dose.
  b. in small multiple doses.
  c. to a small area.
  d. with a filter.

184. Which of the following is a property of radiation that accounts for its effect on biological systems?
  a. It has different energies.
  b. It has no electric charge.
  c. It causes ionization in matter.
  d. It travels at the speed of light.

185. The degree to which recovery is possible after exposure to radiation is related to
  a. total dose received.
  b. degree of protraction.
  c. LET of the radiation.
  d. all of the above.

186. A piece of paper will provide adequate protection from which of the following types of radiation?
  a. alpha
  b. beta
  c. gamma
  d. x-ray

187. LET stands for
  a. lethal effective tolerance.
  b. lethal exchange table.
  c. linear energy transmutation.
  d. linear energy transfer.

188. Human cells are divided into two classifications. There are germ cells and
  a. organic cells.
  b. benign cells.
  c. somatic cells.
  d. reproductive cells.

189. What is the process of cell division of somatic cells called?
  a. meiosis
  b. mitosis
  c. synthesis
  d. multiplication

190. The direct hit theory of cell irradiation can be described by which of the following?
  a. The DNA molecule is struck.
  b. The cell nucleus is struck.
  c. The cell cytoplasm is struck.
  d. The cell is ionized.

191. Which of the following would describe the shape of a DNA molecule?
a. elliptical
b. hypocycloidal
c. double helix
d. circular

192. Which of the following is a type of damage that can occur to a DNA molecule when exposed to radiation?
a. change in the genetic code
b. breakage of the chromosomes
c. breakage of the DNA molecule
d. all of the above

193. What do the letters DNA stand for?
a. direct nuclear action
b. deuteron nucleus activity
c. deoxyribonucleic acid
d. deoxyribose neutral acid

194. How will oxygen retention affect the radiosensitivity of a cell?
a. Radiosensitivity will increase.
b. Radiosensitivity will decrease.
c. Radiosensitivity will be eliminated.
d. Radiosensitivity will not be affected.

195. Which term describes the separation of water into hydrogen and oxygen due to irradiation?
a. radioactivation
b. radiolysis
c. interaction
d. toxicity

196. Which of the following types of patients/people are most radiosensitive?
a. fetal
b. pediatric
c. adult
d. geriatric

197. What is meant by an indirect effect of ionizing radiation?
a. A genetic effect is produced.
b. Ionization in one location may produce an effect in another.
c. An organism other than the one irradiated may be effected.
d. Potentially lethal metabolic dysfunction is produced.

198. Why is free radical formation in the human body so dangerous?
a. Scatter radiation is produced in the body.
b. All radiation shielding is penetrated.
c. Carcinogenic tendencies have been proven.
d. Toxic effects have been observed.

199. Which of the following will improve or will lower when employing beam-limiting devices?

1. image quality
2. patient dose
3. production of scatter radiation

a. 1 only
b. 2 and 3 only
c. 1, 2, and 3
d. none of the above

200. Adding or increasing filtration will primarily protect which of the following?
a. the radiographer
b. the reproductive organs
c. the patient's skin
d. the patient's bone marrow

Radiography Practice Exam I Answers

Patient Care and Education

1. c. Airborne isolation requires anyone entering the patient's room to wear a mask and gown for protective purposes. Gloves are a universal precaution and are required for all examinations.

2. c. When performing two-person CPR the ratio is 15:2.

3. b. The patient must be NPO for 8–10 hours prior to the procedure.

4. b. Tachycardia is a pulse of more than 100 bpm.

5. b. It increases the attenuation and decreases the density.

6. d. Diastolic is the pressure of the heart during relaxation.

7. c. Dyspnea is when a patient is having difficulty breathing.

8. c. Anybody with low blood sugar is considered hypoglycemic.

9. b. Vector transmission occurs when animals transmit infectious organisms such as mosquitoes and ticks.

10. a. An advanced directive should be filled out by the patient in case he or she becomes incapacitated.

11. d. Name, date, medical record number, and anatomical markers should be present on a radiograph.

12. b. An IV bag must be kept at a minimum of 18 inches above the site.

13. c. Patients tolerate the mixture better at 100°F.

14. d. Systolic pressure is the measurement of the heart during contraction.

15. a. A direct injection into the duct is required during a surgical cholangiogram.

16. b. Hypovolemic shock is when the patient experiences a severe loss of blood or plasma.

17. a. Air is considered to be a positive contrast agent. A great example of positive contrast is a chest radiograph.

18. d. A delayed reaction to iodinated contrast could take place up until 24 hours postinjection.

19. b. The complete removal of organisms from equipment and environment is called surgical asepsis.

20. d. They decrease the attenuation and increase the density.

21. c. Aqueous iodine compounds, such as gastrografin, are used when an ulcer or ruptured appendix is present.

22. a. Ionic contrasts are associated with a higher risk of reactions.

23. a. An anaphylactic reaction can cause nausea, vomiting, hives, flushed skin, or fever.

24. d. A ventilator is a mechanical breathing device that assists patients who cannot breathe on their own.

25. a. A compound fracture is presented when the affected bone (or fragment of bone) breaks and protrudes through the surface of the skin.

26. c. Based on contrast type, peristalsis, rate of transmission, and method of voiding/evacuating, exam order should be plain diagnostic film, voiding (IVP, VCU cystogram), evacuating (BE), and drinking study (UGI).

Equipment Operation and Maintenance

27. c. The capacitor discharge portable machine discharges all its stored energy and must be charged after each usage.

28. a. Adherent scatter is not another name attributed to coherent scatter.

29. c. The motor utilized to power the spin of the anode is an induction motor.

30. a. Pair production occurs higher than 2.4 MeV, which is out of the diagnostic range.

31. a. The focusing cup uses a slightly more powerful negative charge than the electrons emitted by the cathode in order to keep them together.

32. **c.** The backup timer should be set to terminate the exposure when 150% of the exposure saturates the chamber.

33. **b.** The acronym ABC stands for automatic brightness control.

34. **a.** The input phosphor is the first aspect of the fluoroscopic tube that receives the incoming remnant photons.

35. **b.** A projectile electron is one of the byproducts, along with the cascade effect of characteristic interactions.

36. **c.** The cathode has a negative charge and the anode has a positive charge.

37. **c.** The acronym AEC stands for automatic exposure control.

38. **b.** Capacitor discharge is most associated with older mobile units.

39. **c.** The kilo-equivalent method is not one of the quality assurance tests required.

40. **d.** The dead man switch is the most popular exposure switch that will automatically terminate the exposure if no pressure is physically being placed on it.

41. **b.** The output phosphor emits light.

42. **b.** *Bremsstrahlung* means braking, and it is a braking method of x-ray production.

43. **d.** Air is utilized within the ionization chambers in an AEC unit.

44. **d.** The anode is commonly a tungsten-rhenium compound.

45. **b.** The photocathode serves to convert light into electrons.

46. **c.** A conductor allows for the free flow of electrons.

47. **d.** All are necessary for the proper warm-up.

48. **b.** Remnant photons interact with the input phosphor of the image intensification tube of a fluoroscopic unit.

49. **d.** The photocathode is present in the image intensifier.

50. **c.** The focusing cup is attributed to the anode.

Image Production and Evaluation

51. **c.** Magnification is effectively reduced by having the least amount of object-to-image distance possible.

52. **c.** The use of the direct square law ($\frac{mA_1}{mA_2} = \frac{D_2^2}{D_1^2}$) allows the technologist to compensate mA for increases in distance.

53. **b.** The largest focal spot size, the longest OID, and the shortest SID all contribute to focal spot blur.

54. **a.** The 15% rule states that a 15% increase in kVp will effectively double the density on a film.

55. **b.** Increasing photon energy by increasing kVp will decrease contrast on a film.

56. **c.** The high speeds of rare earth screens allow for lesser exposures for patients.

57. **d.** Scattered radiation is produced by the patient after attenuation.

58. **c.** $40 \text{ mA} = x \times 0.08 \text{ sec}, x = 500 \text{ mA}$.

59. **c.** MF = SID/SOD, MF = 60/52, MF = 1.15.

60. **a.** As mAs decreases, density is decreased proportionally.

61. **b.** New mAs = old mAs × (new GCF/old GCF), $x = 32 \times (3/6), x = 16$.

62. **c.** The Kodak GBX filter is the most commonly associated safelight filter with rare earth screens.

63. **c.** The acidic properties of the fixer neutralize the alkaline developer.

64. **c.** The smaller the focal spot, the better the detail. The focal spot can only image an object of its size or larger.

65. **d.** Bone has the highest attenuation properties.

66. **a.** This is the definition of the half value layer.

67. **c.** Polyester is the material from which the base of film is constructed, due to its physical properties.

68. **b.** Due to the large increases in exposure necessary to produce a diagnostic image with direct exposure film, it is not commonly used.

69. a. The high degree of scatter cleanup is the main advantage of high ratio grids.

70. d. All are classified as rare earth phosphors.

71. d. All factors (high speeds, high kVp, low mA) contribute to quantum mottle.

72. a. This is the definition of density.

73. d. mA, time, and kVp all can be used in order to regulate density.

74. b. mAs = mA × time, 32 mAs = 600 mA × x, x = 0.05 sec.

75. c. mAs = mA × time, 60 mAs = x × 0.10 sec, x = 600 mA.

76. a. Short-scale contrast refers to a small amount of densities present, which have great differences in their appearances.

77. a. Increasing the kVp will decrease contrast on a film as well as increase density.

78. a. As film speed increases, the density on a film will increase as well, therefore lower mAs should be utilized to compensate.

79. c. Filtration's primary purpose is to decrease skin dosage by eliminating the majority of nondiagnostic low energy photons.

80. a. Using the 15% rule to double density by increasing kVp, the mAs must be halved in order to compensate for the difference, thus decreasing dose.

81. c. Due to the inverse square law, doubling distance reduces density by a factor of 4.

82. c. MF = SID/SOD, x = 44/36, x = 1.2 (1.16)

83. d. New mAs = old mAs × (old screen speed/ new screen speed), x = 5 × (800/100), x = 40 mAs.

84. a. Contrast will increase due to restriction of the beam by limiting the area with which the beam will interact.

85. b. Pneumonia is an additive pathology. Bacteria produce fluid in the lungs, thereby requiring an increase in the technique used.

86. b. Film should be stored between 40% and 60% humidity.

87. a. Quantity of photons produced refers to mAs.

88. a. A decrease in photon energy will lead to a higher contrast film.

89. b. This describes the 15% rule.

90. d. The analog to digital converter converts the light from the phosphor into a matrix.

91. b. 400 mA is required with 0.25 seconds to create 100 mAs.

92. d. 40 mAs.

93. d. The highest mA (Choice D, 600 mA) produces the greatest density.

94. a. x = 2.5 × (2/1), x = 5.

95. c. The lowest possible exposure time is the best way to combat involuntary motion.

96. b. To compensate for an increase of 15% in kVp (which will double density), mAs should be reduced by half of its original value.

97. b. All of these contribute to shape distortion.

98. d. Motion will lead to the greatest degradation of recorded detail.

99. d. Tube angulations lead to shape distortion.

100. b. Aluminum is the most common filtering material.

Radiographic Procedures

101. a. The 30 degree medial oblique of the foot best demonstrates the fifth metatarsal tuberosity.

102. c. There are 33 vertebrae in the spinal column (7 cervical, 12 thoracic, 5 lumbar, 5 fused sacral, 4 fused coccyx).

103. c. The PA oblique projection of the wrist best demonstrates the lateral carpal bones: the scaphoid, the trapezoid, and the trapezium.

104. c. The right SI joint is best demonstrated utilizing the RAO (or LPO) projection.

105. b. The AP oblique projections (LPO/RPO) of the lumbar spine are used to best demonstrate the zygapophyseal articulations.

106. a. The midcoronal plane separates the body into equal anterior and posterior halves.

107. c. The 35–45 degree posterior oblique of the shoulder (Grashey) is used to demonstrate the glenoid cavity.

108. b. The swimmer's position is best used for trauma/broad shouldered patients in order to best visualize the lower cervical/upper thoracic vertebrae.

109. c. There should be ten posterior ribs visualized in a properly exposed PA chest x-ray done upon the second full inspiration.

110. b. For an AP projection of the sacrum, the central ray is angled 15 degrees cephalad directed 2.5 inches superior to the symphysis pubis.

111. d. The capitate is the largest carpal bone.

112. d. Chest projections are performed at 72 inches in order to minimize heart magnification.

113. c. The left lateral decubitis with the patient lying on the left side with a horizontal center ray directed at the level of iliac crest, as well as the upright projection with the upper margin of the IR above the level of the diaphragm, are both best for demonstrating free air in the abdomen.

114. a. The most frequently fractured site of the shoulder is the surgical neck, which is a much thinned out section of the humerus.

115. b. The mastoid tip and the lower edge of the upper incisors need to be perpendicular to the film in order for the dens to be visualized through the open mouth.

116. a. The LAO position for ribs best demonstrates the right axillary ribs, which is the side furthest from the IR. The RPO position also demonstrates the right axillary ribs.

117. b. The zygapophyseal joint is best visualized in the lateral view of the cervical spine.

118. a. The scaphoid (navicular) is the most frequently fractured carpal.

119. c. The Clements-Nakayama Method is best utilized in order to image the hip and proximal femur if bilateral hip fractures are suspected. It angles the center ray and the film both to the femoral neck and posteriorly in order to visualize the hip in a lateral position.

120. b. The Settegast (or sunrise) Projection best demonstrates the patella free of superimposition as well as the patellofemoral joint.

121. b. The midsagittal plane separates the body into equal right and left halves.

122. a. The Fuchs Method projects the dens through the foramen magnum.

123. a. The AP oblique LPO position will demonstrate the fundus and duodenal bulb opacified by positive contrast media while the pylorus and body will be filled with air.

124. c. The tibial plateau is palpable on the anterior proximal tibia.

125. c. The stomach and splenic flexure of the colon are located in the left upper quadrant of the abdomen.

126. c. The PA axial Camp Coventry Method and the AP axial Holmblad Method of the knee both allow for visualization of the open intercondylar fossa.

127. d. The catheter should be inserted into the rectum no more than 3.5–4 inches.

128. b. The AP internal rotation demonstrates the lesser tubercle of the humeral head in profile medially.

129. d. The splenic flexure is located at the junction between the transverse and descending colon.

130. d. The lateral projection demonstrates all four paranasal sinuses.

131. d. The apex is the area of the lung lying over the clavicles.

132. c. There are 206 bones in the human body.

133. c. Bisect a line between ASIS and symphysis pubis and drop inferiorly and perpendicularly one inch for hip localization.

134. c. In a properly positioned Rheese Projection, the optic foramen is visualized in the lower lateral quadrant of the orbit.

135. a. It is imperative that all sinus studies be performed erect in order to properly differentiate air/fluid levels within the paranasal sinuses.

136. b. The small intestine is the primary site of nutrient absorption.

137. a. The ureter is the connection from the kidneys to the bladder. Urine is transported down.

138. a. There are five lumbar vertebrae.

139. a. The external soft tissue portion of the ear is referred to as the auricle.

140. b. The LPO projection of a barium enema series will allow the splenic flexure to be visualized with limited superimposition of the adjacent areas of the colon.

141. a. The vomer is one of the 14 facial bones, not a cranial bone.

142. b. Axis refers to C-2, which allows the head to pivot left and right.

143. c. If IOML is used for the AP axial Townes Method for the cranium then a 37 degree central ray is utilized; for OML it is 30 degrees.

144. c. The condylar process articulates with the temporal bone at the TMJ.

145. b. The t-spine has a kyphotic curve.

146. b. The Stecher Method, which is a PA projection in ulnar flexion with the fingers elevated 20 degrees from the plane of the IR, is best to visualize the navicular.

147. d. Sthenic patients make up approximately 50% of the population.

148. c. The odontoid process, or dens, extends vertically from the body at C-2 and articulates with C-1. It helps to provide the pivot for the head.

149. a. Peristalsis is strongest in the beginning of the digestive process, which starts after swallowing in the esophagus and proceeds through the stomach to the small bowel and finally to the colon where peristalsis is the slowest.

150. b. The orbitomeatal line (OML) refers to the baseline drawn between the outer canthus of the eye and the EAM and is utilized in positioning of the skull.

151. a. This describes the Grashey position for demonstration of the glenoid fossa.

152. b. IPL should be perpendicular to the IR in a lateral skull projection.

153. b. The femur is a long bone.

154. c. The AP axial Townes Projection best demonstrates the occipital bone due to its 30 degree caudal angle, which frees it from superimposition.

155. c. For maxiallary sinuses, the Waters Projection, a PA projection utilizing a 37 degree angle between the tip of the nose and the IR with the chin resting on the upright grid device with the CR exiting the acanthion, is best.

156. d. The three aspects of the small intestine in order, starting from the stomach are duodenum, jejunum, and ilium.

157. c. For the lower ribs, the exposure should be made after exhalation in order to utilize the diaphragm as contrast.

158. b. In the LAO or RPO positions for SI joints, the left SI joint is the part of interest.

159. c. For a parietoacanthial projection (Waters Method) of the sinuses, the orbitomeatal line should form a 37 degree angle to the plane of the IR. This is done by tilting the head back until this angle is attained.

160. a. The enema bag should be placed 18–24 inches above the anus in order to facilitate the flow of barium and best opacify the colon.

Radiation Protection

161. d. Intensity has nothing to do with protective barrier thicknesses.

162. c. Ten or more inches are acceptable.

163. d. All choices are reasons for repeating a radiograph.

164. c. Leakage is an issue for the radiologic technnologist.

165. c. The grid is the only one not located between the patient and the x-ray tube.

166. d. A shadow shield is suspended between the patient and the tube.

167. c. Any time x-rays pass through matter, scatter occurs.

168. b. GSD in the general population cannot exceed 1 rem.

169. d. GSD for the occupational population cannot exceed 5 rems.

170. a. Distance is the primary factor.

171. c. This is the definition of skin erythema.

172. a. Rem is used as the unit of measure for the occupational dose. It stands for roentgen equivalent man.

173. b. Low WBC is a quick response to radiation.

174. a. Collimating is the best way to lower exposure to matter.

175. a. Lymphocytes are the least radioresistant.

176. c. Filters allow radiation energies to be identified.

177. a. Thermo luminescent dosimeters (TLD) are heated to be read.

178. b. The shoulder is farthest from the reproductive organs.

179. d. High phase and high kVp will produce the lowest patient dose on a spine radiograph.

180. b. Sievert is a measurement of dose equivalence.

181. c. The lowest grid will provide the lowest dose to a patient's skull.

182. b. The size of the cell does not influence radiation effect.

183. a. One dose delivery causes the most profound effect.

184. c. Ionization causes the effect on biological systems.

185. d. All of the above matter to recovery.

186. a. Alpha has high LET, but it lacks penetrability.

187. d. The acronym LET stands for linear energy transfer.

188. c. The two main classes of human cells are somatic cells and germ cells.

189. b. Mitosis division is related to somatic cells.

190. a. The DNA molecule is struck in the direct hit theory.

191. c. The DNA molecule is in the shape of a double helix.

192. d. All of the above can occur to a DNA molecule when exposed to radiation.

193. c. DNA stands for deoxyribonucleic acid.

194. a. Oxygen enhances radiosensitivity.

195. b. Radiolysis is the separation of water into hydrogen and oxygen due to irradiation.

196. a. The younger the person, the more radiosensitive he or she is.

197. b. An indirect effect happens in another area of the body.

198. d. Toxicity is related to free radical formation.

199. c. Beam-limiting devices improve all three factors mentioned.

200. c. Patient skin exposure is improved with filtration.

▶ RADIATION PROTECTION

CHAPTER SUMMARY

Radiation protection is one of five sections on the ARRT examination, and you must learn about it in order to attain state licensure and national certification in radiography. The following pages contain information as a review for this section, followed by multiple-choice questions. Each of the following topics can be found on the list of content specified by ARRT, and it is information covered in accredited radiography programs.

Key Words

automatic exposure control (AEC)	milli
ALARA	nonstochastic effect
attenuation	primary beam
beam-limiting devices	radiation absorbed dose (RAD)
gray (Gy)	relative biological effectiveness (RBE)
image receptors	rem
lethal dose (LD)	remnant radiation
leakage radiation	roentgen (R)
linear energy transfer (LET)	secondary beam

Radiation Exposure and Protection

Dose-response relationships: These are dose-to-response results projected on a graph or chart. Results are demonstrated as either linear or nonlinear, and are either threshold or nonthreshold. Any observed measurement stated as linear means that the response is directly proportional to the dose. In other words, the dose equals the response. In nonlinear situations, this is not the case, and the dose response would be anything but proportional. Threshold relationships mean that there are doses that do not demonstrate an effect, and nonthreshold-dose relationships imply that an exposure amount would cause or produce an effect. These relationships are demonstrated in the following chart.

Threshold Relationships

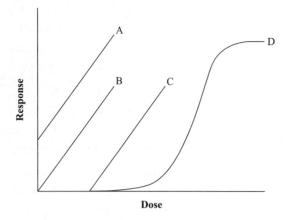

A = linear, nonthreshold; B = linear, nonthreshold; C = linear, threshold; and D = nonlinear, nonthreshold (more specifically: sigmoid threshold, because of the S-shaped line).

Stochastic effects: Stochastic effects are random in nature. This means that out of a group of individuals who are irradiated, some will develop an effect, such as cancer or an abnormal genetic effect, and some will not.

Nonstochastic effect: The best example of this would be if everyone in a group were to be irradiated, everyone would develop an effect, such as cataracts or blood changes (lymphocyte reduction).

Linear energy transfer (LET): As energy is transferred into and passes through soft tissue, it will deposit its energy into that soft tissue. The rate at which it does this is called linear energy transfer.

Relative biological effectiveness (RBE): RBE is the comparative measurement of different types of radiation compared with x- and gamma-rays. LET and RBE are directly proportional. As LET goes up, so does RBE, and vice versa. In addition, as LET of radiation increases, so will biological damage.

Lethal dose (LD): You will always see a fractionation with the LD designation. It will indicate the percentage of the population that will die in a certain time period. For example, LD 50/30 means that provided a specific population receives the same dose, 50% of that population will die in 30 days from exposure. There are progressive biological effects to the body, depending on the strength of the dose. Starting with approximately 20 rem, the body may show an effect. The effect would start with blood changes, and as the dose increased, the biological effects would be nausea, diarrhea (125–150 rem), erythema, and temporary sterility. At an LD 50/30, the effect can be death (approximately 600 rem or more). General types of cell damage include somatic effects, as well as genetic effects. Somatic effects encompass damage to the body, including the cell, and genetic effects involve damage to only the cell's genetic portion, which is contained in the DNA.

Somatic Effects

Somatic effects usually occur when there is a large area of the body that is exposed. Obviously, the larger the dose to a larger part of the body, with higher LET exposure, would increase the effect. In addition, the doses being mentioned here are of much higher consideration than that of the diagnostic exposure range.

Examples of early somatic effects include hematopoietic syndrome (blood changes), gastro-intestinal effects, and effects to the central nervous system. If any of these effects is high enough, death can occur. Again, this would be based on high doses of radiation, exceeding the diagnostic range, and the exposure would involve exposure to a large area of the body. The exposures would be so large that the onset of the effect of the exposure would be fast.

Examples of late somatic effects would be of the long-term kind. Certain types of cancer would be examples of this. In addition, cataracts, embryologic effects, thyroid complications, and shortening of the overall life span may occur. It should be noted that these effects are caused by doses outside of the diagnostic range.

Genetic Effects

Reproductive (germ) cells are also affected by doses of radiation, with differences between male and female reproductive cells. Male (testes) include both mature and immature cells. Male mature cells are essentially radioresistant, and the immature male germ cells are very radiosensitive. Because of this, small doses of radiation may cause periods of temporary sterility, while larger doses can cause long-term or permanent sterility. It should be noted that the dose levels that cause these type of effects almost never happen at the diagnostic level because the dosing for diagnostic exams are so low. In females, the ova do not divide regularly. Hundreds of mature ova are produced by a women in her childbearing life, with one egg being passed down to the uterus every 30 days or so.

Immature ova are radiosensitive and mature ova are not. Ova can carry exposure-damaged chromosome(s) and can still unite with a sperm cell, with the result being genetically damaged offspring, or genetically damaged offspring in future generations. In addition to this, exposure to ionizing radiation can also cause sterility (when new and mature ova are exposed). Even though most exposures which carry the risks mentioned in this section are doses that exceed the diagnostic range, shielding with lead-made devices during diagnostic studies should be practiced on all types of patient types, especially young patients and ones of childbearing age. Of course, shielding cannot always be applied, but it should be practiced whenever possible.

Exposure and Protection

Photon interactions with matter must be stressed when exploring the basics of radiation protection. There are several definitions of these interactions. In the Compton effect, the scattering of x-rays results in ionization and loss of energy. Scatter radiation occurs any time x-rays traverse through matter. Photoelectric absorption is the absorption or energy transfer of x-rays by the process of ionization.

Coherent (classical) scatter refers to the scattering of x-rays with no loss of energy. Also referred to as Rayleigh or Thompson scattering, attenuation is the reduction of intensification as a result of absorption and scattering. Absorption is influenced by the thickness of the body part, and the type of tissue (atomic number). The thicker the body part and the higher the atomic number, the more absorption and the higher the production of scatter radiation.

Low mAs, high kVp techniques reduce exposure to the patient. The 15% rule states that if you reduce the mAs by one-half and increase the kVp by 15%, you can reduce exposure and maintain density. What must be remembered here is that you cannot compensate a kilovoltage level for an inadequate level of mAs, so do not abuse the 15% rule. To achieve a proper balance using the 15% rule requires an acceptable level of mA and kVp. Remember, mAs is the quantity of radiation and exposure, and kVp is the quality of the beam. It is mA that is responsible for the exposure.

Shielding

Shielding is placing lead in the way or over a body part to prevent it from interacting with ionizing radiation. Two main types of lead shielding are the direct contact shield and the shadow shield. The contact shield is placed directly on the patient, usually on an area of the body in close proximity to the area of interest, *but never on the area of interest*. The shadow shield does not come in contact with the patient. It is usually attached to the tube housing and placed in the path of the primary field. The shadow shield is very useful in the operating room, where the sterile field is always a factor. The shaped contact shield is another type of contact shield, almost always used for gonad shielding for males. For females, the fig leaf shield is recommended. It should be noted that depending on what area is being examined, it may not be practical to shield a body part. You would not want to use a shield over an area of interest, as it would defeat the purpose of radiographing that body part.

Total Filters

The use of proper filtration results in lower exposure to skin and internal body organs. The basic principle of a filter is to allow the entire primary beam to pass through it, and it removes the lower energy photons that are not useful, making the beam harder and more homogenous. Filters will remove weaker photons (by absorption), which will improve the effect on average beam energy. A filter is usually placed below the anode/cathode region of the tube and above the patient. In most cases, it is placed within the tube housing.

There are compensating filters, such as wedge and trough filters, which are primarily used when the thickness of the body part being irradiated changes drastically from one end to the other, such as with the foot.

Automatic Exposure Control (AEC)

AEC, when used correctly, is a genuine radiation-reducing technique. An ionization chamber is placed behind the patient, and it will actively terminate the exposure when the correct and sufficient amount of radiation has passed through the patient. It is of great importance for the technologist to position the patient correctly so the exposure terminates at the proper time. Patients with a lower thickness (thinner) can get too much exposure due to minimum response time. Even when performing x-ray examinations without the use of AEC, proper positioning is important for exposure reduction because it reduces the need for repeat exposures. Also, the practice of good patient communication is important in radiation protection. Proper communication helps to relax the patient, and therefore will improve the cooperation the patient gives the technologist during the procedure. A relaxed patient is a cooperative patient.

Image Receptors

Image receptors are also important to understand. Remember, the higher the intensifying screen number, the lower the mAs to achieve the appropriate density. Grids are used primarily to improve the visibility of detail of the image, but how are they a factor in radiation protection? Less radiation to the patient is a result of the use of the lowest grid ratio. In addition, if the image is achieved with a grid, it helps to reduce the need for a repeat image. It is mentioned over and over, but one of the most important aspects of radiation protection is reducing the repeat radiograph. The use of grids significantly increases the amount of exposure to the patient, sometimes up to ten times more than without a grid. Using the image intensifier mode helps to reduce the amount of exposure to the patient. The image is reduced in size, and the smaller area of the image intensifier is utilized.

The Cardinal Principles of Radiation Protection

There are three cardinal principles to radiation protection: time, distance, and shielding.

1. **Time** (reduction of irradiation time): This is mostly a fluoroscopic issue. If/when the physician reduces the time of the fluoroscopic portion of the exam, irradiation (exposure) is also reduced. The dead man switch (a foot pedal that activates the fluoroscopic beam), as well as the audible fluoroscopic timer (which informs the fluoroscopist of specific time frames of exposure) are helpful tools.

2. **Distance:** Increase the distance from the source, which is the space between the source and technologist or medical care provider and the patient.

   Inverse Square Law: The diagram that follows helps to explain the inverse square law, which indicates that the distance from the source is inversely proportional to the distance of the square. If you move away from the source (doubling the distance), you reduce the intensity of the beam to one-quarter of what it was. The inverse square law is represented mathematically as seen in the diagram. The very top of the pyramid represents the x-ray tube. The distance d1 would be a specific source-to-image distance (SID), such as 40 inches, and the distance d2 would be about double that of d1, 80 inches. The x-ray beam is demonstrated by the dotted line, so you can see that if the distance of the SID is doubled, the same amount of beam is spread out over an area that is four times as large, and each part of the beam is one-fourth of the intensity. So, the field E, F, G, H is four times larger and one-fourth as intense as field A, B, C, D.

Inverse Square Law

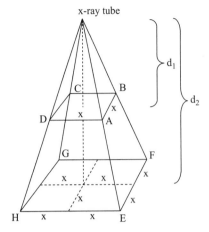

Source-to-image distance (SID) would be one way to reduce exposure, but specifically in fluoroscopy, increasing the distance between the patient and the x-ray source will also reduce exposure to the patient.

The inverse square law is represented mathematically as:

$$\frac{I_1}{I_2} = \frac{d_2^{\,2}}{d_1^{\,2}}$$

Where I_1 = Original distance exposure intensity
I_2 = new distance exposure intensity
d_1 = original distance from source
d_2 = new distance from source

3. **Shielding** (use lead as the type of barrier): Lead is used as the barrier for protection against x-radiation. Devices of radiation shielding made of lead include protective drapes, protective bucky slot covers, aprons, gloves, thyroid shields, and glasses (glass is lead tinted). Lead aprons have been in use protecting technologists, staff, and doctors since x-rays were first discovered and it was realized that excessive amounts of them are dangerous. Full-body aprons help reduce exposure to bone marrow areas of the body by up to 80%, and they are the main type of shielding available.

The four main types of shields are:

1. Flat contact shields
2. Shadow shields
3. Shaped contact shields
4. Clear lead shields

ALARA: ALARA stands for *as low as reasonably achievable*. It is a way of thought, and it means that the technologist performing the radiological scan will do everything he or she can to minimize the effects of radiation to self and patient. When applying the cardinal principles (time, distance, shielding), you are in effect activating the ALARA principle.

Dosimeters

Dosimeters are devices that monitor occupational exposure. They should be worn only when in the occupational setting. Types of dosimeters include the following: film badges (most popular), extremity dosimeters (ring badge, used in nuclear medicine), pocket ionization chamber (immediate reading available), and thermo luminescent dosimeter (TLD, lithium fluoride). These devices may be worn with a lead apron, and the dosimeter should be worn *outside of the apron*, at the collar on the anterior of the body.

Please note: Dosimeter devices in no way protect one from exposure. They are devices which inform "after the act." That is, they provide you and your institution with an idea of how well you practice good radiation protection techniques. Your personal records are cumulative and follow you from one job to another. A summary of National Council for Radiation Protection (NCRP) recommendations are discussed in the next section.

Federal Standards

The federal government has standards that are enforced by the National Council for Radiation Protection. The NCRP collects and analyzes data, which are used to protect people (both the general and occupational populations) against overexposure. Some of these standards involve limits on total exposure allowable for different classifications of the general public, including radiation workers. For example, the NCRP permits an annual effective dose of 50 millisieverts (mSv), or 5 rem, for the occupational population. This is a higher amount than is allowed for the general public, which is 1 mSv (0.1 rem).

You may want to know how the dose can be higher for radiation workers. First, you must be aware of the genetically significant dose (GSD). The GSD is the average annual gonadal dose allowable for members of the general population who are of childbearing age. The number of radiation workers in the general public is not a large enough group to affect the potential overall genetic effects in the United States. Because of this, it is really the responsibility of the occupational worker to practice ALARA principles so the potential of somatic and genetic damage is kept as low as possible.

Please see NCRP Report No. 116, Limitation of Exposure to Ionizing Radiation. NCRP Report No. 116 is the latest in a long series of reports on basic radiation protection criteria that began in 1934. The current report takes advantage of new information, evaluations, and thinking that have developed since 1987, particularly the risk estimate formulations set out in NCRP Report No. 115. While the recommendations set out in this report do not constitute a radical revision of the basic criteria, they do represent a refinement of the system enunciated in Report No. 91. Important changes include the utilization of revised tissue/organ weighting factors and the introduction of radiation weighting factors. Also noteworthy is the introduction of an allowable reference level of intake. Additionally, there is a recommendation of an age-based lifetime limit for occupational exposure and a major simplification of limits aimed at controlling the exposure to the embryo and fetus. This report, after outlining the goals and philosophy of radiation protection and exposure limits, goes on to review absorbed dose, equivalent dose, effective dose, radiation weighting factors, guidance for emergency occupational exposure, and remedial action levels for naturally occurring radiation.

Natural Sources of Radiation Exposure

The earth's environment has always had natural radiological sources. When learning about the radiologic sciences, we get caught up in manmade sources (focusing on medical exams). However, to put it bluntly, we are being irradiated by natural sources all the time. Natural sources include, but are not limited to: radon exposure, cosmic radiation, terrestrial radiation, and internal sources.

Sources	Definition
Radon	From radium in the earth's surface
Cosmic	Exposure from outer space—pilots, beware!
Terrestrial	From the earth, metals, buildings
Internal	From within the human body

Example: A United States resident, on the average, is exposed to approximately 198 mrems per year of carbon-14.

Please see the following chart for more information.

Natural Sources of Radiation Exposure

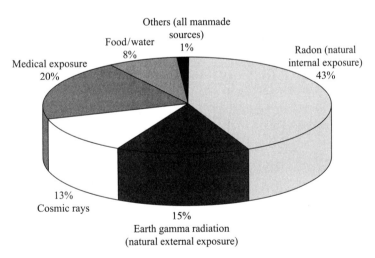

Others (all manmade sources) 1%

Food/water 8%

Medical exposure 20%

Radon (natural internal exposure) 43%

13% Cosmic rays

15% Earth gamma radiation (natural external exposure)

The Heterogenous Beam

The following diagram helps to explain the four types of photons that make up the heterogeneous beam. Primary photons (1, 2, 3, and 4) emerge from the x-ray source. Remnant photons (1 and 2) pass through the object being radiographed and reach the film. Attenuated photons (3 and 4), having interacted with atoms of the object, are scattered or absorbed and do not reach the film.

Four Types of Protons
That Make Up the Heterogeneous Beam

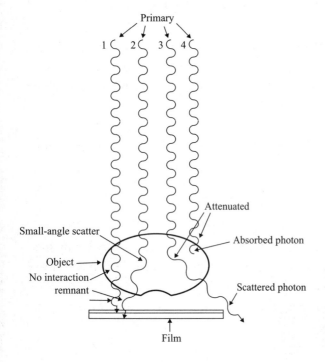

Fluoroscopic Equipment

Fluoroscopy is the diagnostic procedure that produces the highest levels of radiation exposure. Fluoroscopic equipment uses continuous real-time imaging, so it is unlike regular diagnostic radiography. The physician usually does fluoro, so one role of the radiographer is to monitor and remind the doctor of the exposure length to the patient. One of the more important tools in fluoroscopy is the dead man switch. This switch, usually located on the floor, must be stepped on to activate the exposure. This helps to trim down the exposure time. It is called dead man because if the fluoroscopist were to fall down or collapse, the exposure would terminate. Another unique piece of equipment in fluoroscopy is the protective curtain. This curtain hangs between the patient and the fluoroscopy operator. There is also a timer, which will terminate exposure after a desired set time, usually five minutes. The following image shows fluoroscopic equipment, including a lead apron and protective curtain.

Fluoroscopic Equipment Operation

Portable X-Ray Equipment

Like fluoroscopy, portable imaging has some differences worth noting. Lead aprons should be available and worn if mobile barriers are unavailable. Just about all mobile exposures are done outside the x-ray department, so there are usually no protective barriers.

Points to be noted:

- Lead aprons should be worn if mobile barriers are unavailable.
- Least scatter is at a 90 degree angle from the patient.
- Utilize the inverse square law to reduce dose (use exposure cord at full length).
- The cord on the exposure switch must be at least 6 feet long.
- Radiographer should never hold the cassette during exposure.

Definitions

Attenuation: Any process that decreases the intensity of the beam that was directed toward some destination.

Beam-limiting devices: Devices (located between anode and cathode and the patient) which restrict the size of the field being irradiated. Cones and collimators are prime examples. Other than limiting the field, these devices improve visibility of detail by eliminated photons for prevention of potential scatter radiation.

Gray (Gy): International standard for the absorbed dose; equal to the RAD.

Leakage radiation: Radiation which escapes from the tube housing, usually traveling in a different direction than that of the primary beam. Remember that when x-rays are created within the tube, the direction they travel is of an isotrophic nature (in all directions).

Milli: Prefix which stands for one-thousandth; for example, a milligray = one-thousandth of a gray.

Primary beam: The beam as it travels out of the tube and housing, up until the point when it interacts with living tissue.

RAD: Radiation absorbed dose.

Rem: Roentgen equivalent man.

Remnant radiation: All of the x-ray photons that reach their destination (the film) after passing through the object being radiographed; the photons that have contributed to the image.

Roentgen (R): Named for the discoverer of x-radiation, roentgen is also the term used for the measurement of exposure, 2.58 x 10–4 coulombs per kilogram of dry air.

Secondary beam: As soon as the primary beam interacts with the living tissue, it is identified as secondary.

Sievert (Sv): International standard for the rem (dose equivalent).

Scatter radiation: The process in which the x-ray photons undergo a change of direction after interacting with the atoms of an object. This is the type of radiation that the operator is most concerned with as far as exposure is concerned.

Chapter 5 Review Questions

1. A radiation-caused effect that appears in the offspring of the exposed parent is termed a(n)
   a. acute mutation.
   b. long-term mutation.
   c. genetic mutation.
   d. somatic mutation.

2. The acute radiation syndrome associated with changes to blood and blood-forming body parts is called the
   a. hemopoietic syndrome.
   b. CNS syndrome.
   c. GI syndrome.
   d. prodromal syndrome.

3. A large radiation exposure to the lens area of the eye will most likely result in the formation of
   a. glaucoma.
   b. scleroma.
   c. retinitis.
   d. cataracts.

4. An exposure level of radiation at which a biological response can be noted is termed the
   a. tolerance dose.
   b. ambient dose.
   c. congenital dose.
   d. threshold dose.

5. A common form of cancer that appears to follow a threshold dose relationship is
   a. lung cancer.
   b. skin cancer.
   c. leukemia.
   d. Hodgkin's disease.

6. In male humans, the first biological effects from radion doses of approximately 20–25 cGy are seen in the
   a. blood count.
   b. skin.
   c. sperm count.
   d. eyes.

7. A fluoroscopic exposure switch which is controlled by the fluoroscopist's foot is called
   a. foot switch.
   b. dead man switch.
   c. fluoro switch.
   d. rotor switch.

8. All of the following are rules for mobile radiography EXCEPT
   a. provide gonadal shielding.
   b. label and handle each cassette to avoid repeat radiographs.
   c. never place your hand or any body part within the primary beam.
   d. never bring the mobile portable unit onto an occupied elevator.

9. The lead thickness that reduces the intensity of the radiation beam to 50% is a(n)
   a. HVL.
   b. 50 reducer.
   c. ten desensitization device (TDD).
   d. TVL.

10. Factors which determine protective barrier thicknesses include all of the following EXCEPT
    a. use.
    b. workload.
    c. time occupancy.
    d. intensity.

11. When operating a mobile x-ray unit, which of the following is not an acceptable source to skin distance?
   a. 10 inches
   b. 12 inches
   c. 8 inches
   d. both **a** and **c**

12. Which of the following is grounds for a repeat radiograph?
   a. underexposure of area of interest
   b. poor positioning of patient
   c. mislabeling of the film (patient identification)
   d. all of the above

13. Leakage radiation on a mobile x-ray unit is a concern for
   a. the patient.
   b. the doctor.
   c. the radiologic technologist.
   d. none of the above.

14. Which of the following devices is not located between the patient and the x-ray tube?
   a. primary beam
   b. filtration
   c. grid
   d. collimator

15. Which of the following is not intended to touch or be placed on the patient?
   a. flat contact shield
   b. fig leaf shield
   c. shaped contact shield
   d. shadow shield

16. You can significantly reduce patient exposure by
   a. using the lowest possible screen speed.
   b. adding a radiographic grid.
   c. reducing the amount of filtration.
   d. using the highest screen speed.

17. With all other factors remaining the same, which grid ratio will result in the lowest patient dose?
   a. 8:1
   b. 6:1
   c. 10:1
   d. 16:1

18. Which primary protective barrier height exceeds the necessary height needed?
   a. 7 feet
   b. 8 feet
   c. 6 feet
   d. both **a** and **b**

19. Where is the protective lead curtain located?
   a. in the doorway leading to the control booth
   b. between the patient and fluoroscopist
   c. on the patient
   d. between the physician and the technologist

20. The lead thickness that reduces the intensity of the radiation beam to 10% of its original strength is
   a. TVL.
   b. ten reducer.
   c. ten desensitization device (TDD).
   d. TRL.

21. Dose measurements are basically stated in terms of dose equivalents and are expressed in units of
   a. rem or sievert.
   b. rad or gray.
   c. roentgen or curie.
   d. RBE or QF.

22. Radon is responsible for roughly _____% of a human's average total exposure from natural background radiation sources.
a. 15
b. 25
c. 50
d. 75

23. When mA is increased, how must exposure time be adjusted in order to maintain the same amount of radiation exposure?
a. Time must be increased.
b. Time must be decreased.
c. Time must be kept the same.
d. Time must be matched with kVp.

24. If the SID is cut in half, which of the following changes must be made in order to maintain the same exposure?
a. Exposure must be halved.
b. Exposure must be reduced by four times.
c. Exposure must be doubled.
d. Exposure must be increased by four times.

25. When is scatter radiation created?
a. When a grid is used
b. When you use too much kVp
c. When x-rays pass through matter
d. none of the above

26. The GSD for the general population cannot exceed _____ rem(s).
a. 3
b. 1
c. 2
d. 5

27. The GSD for the occupational population cannot exceed _____ rem(s).
a. 3
b. 1
c. 2
d. 5

28. What is your best protection from scatter when performing a portable radiograph?
a. to stand as far away as possible
b. to collimate to the size of the film
c. to set a higher KV if possible
d. to use high-speed rotation if available

29. Redness of the skin from radiation exposure is referred to as _____.
a. exfoliation
b. purulent
c. erythema
d. erythrocyte

30. Occupational dose (radiation workers) is expressed using what unit of measurement?
a. rem
b. rad
c. LET
d. roentgen

31. How are wavelength and energy of the x-ray beam related?
a. directly
b. inversely
c. chemically
d. empirically

32. The principle function of filtration from the radiographic tube is to reduce
   a. operator dose.
   b. patient skin dose.
   c. image noise.
   d. scattered radiation.

33. Which of the following groups of exposure factors will deliver the least amount of exposure to the patient?
   a. 50 mAs, 100 kVp
   b. 100 mAs, 90 kVp
   c. 200 mAs, 80 kVp
   d. 400 mAs, 70 kVp

34. Which of these exposure types will cause the greatest effect from radiation exposure?
   a. low dose over a long time period
   b. high dose over a short time period
   c. low dose over a short time period
   d. high dose over a long time period

35. Somatic effects of radiation are most related to which of the following choices?
   a. the fetus
   b. the newborn
   c. future generations
   d. the exposed person

36. What does the acronym GSD stand for?
   a. gonadal safe dose
   b. genetically significant dose
   c. genetic-somatic dose
   d. general somatic dose

37. Which area of the body, if irradiated, would affect white blood cells?
   a. lungs
   b. liver
   c. pancreas
   d. bone marrow

38. The primary source of scatter radiation exposure is the
   a. patient.
   b. tabletop.
   c. x-ray tube.
   d. grid.

39. A specific exposure to radiation basically has the greatest potential for damage in tissues with
   a. the largest number of cells.
   b. more rapidly dividing cells.
   c. cells with the highest water content.
   d. the most abundant amount of cells.

40. What is NOT a long-term radiation effect?
   a. leukemia
   b. lowered WBC count
   c. cataracts
   d. life-span shortening

41. Which of the following is the best way to reduce exposure to the reproductive organs of a man?
   a. beam collimation
   b. proper filtration
   c. fast screen speeds
   d. high kilovoltage

42. According to the Law of Bergonie and Tribondeau, which of the following cells are particularly radiosensitive?

   1. Young, immature cells
   2. Stem cells
   3. Highly mitotic cells
   4. Gamma cells

   a. 1 only
   b. 1, 2, and 3
   c. 2, 3, and 4
   d. 1, 2, and 4

43. Why do film badges have filters?
  a. to prevent overheating during exposure
  b. to increase the radiation during exposure
  c. to indicate the type of radiation received
  d. to shield the badge from radiation

44. Which type of dosimeter is heated so it can be read?
  a. TLD
  b. pocket dosimeter
  c. film badge
  d. ionization chamber

45. Of the following different types of personal monitoring devices, which is considered the most accurate in recording dosages?
  a. Film badge
  b. Pocket dosimeter
  c. TLD
  d. None of the above

46. Which technique will produce the lowest patient dose on a spine radiograph?
  a. single phase, 80 kVp
  b. single phase, 65 kVp
  c. three phase, 65 kVp
  d. three phase, 80 kVp

47. Sievert is a known radiation measurement of
  a. exposure.
  b. dose equivalent.
  c. activity.
  d. dose.

48. Which scatter reduction device will require the lowest dose to a patient's skull?
  a. air gap
  b. 8:1 grid
  c. 6:1 grid
  d. 16:1 grid

49. The effects of radiation are known to be influenced by all of the following EXCEPT
  a. dose rate.
  b. size of the cell(s).
  c. type of cell.
  d. type of radiation.

50. The effects of radiation are usually more profound if the
  a. exposure is delivered in one dose.
  b. exposure is delivered in small multiple doses.
  c. exposure is delivered to a small area.
  d. exposure is delivered with a filter.

51. Shorter exposure times, decreased patient dose, and increased latitude in technique are advantages of high
  a. kVp techniques.
  b. contrast techniques.
  c. grid ratio techniques.
  d. mAs techniques.

52. Radiation amounts received during diagnostic studies are chiefly affected by
  a. focal spot size.
  b. patient positioning.
  c. size of field exposed.
  d. type of equipment.

53. Which of the following interactions are most common in tissue at 100 kVp?
  a. coherent
  b. pair production
  c. Compton scatter
  d. photoelectric effect

54. A general purpose radiography unit's minimal filtration thickness is
a. 0.5 mm Al equivalent.
b. 1.75 mm Al equivalent.
c. 2.5 mm Al equivalent.
d. 4.0 mm Al equivalent.

The next two questions refer to the following scenario:

A radiologist stands one foot from the patient while performing fluoroscopy, using a standard under-the-table unit. The scatter to the radiologist is 5 mR/min at 1 foot. The tube output at the tabletop (entrance exposure to the patient) is 2 R/mA/minute.

55. What is the entrance exposure to the patient for 5 minutes of fluoroscopy at 2 mA?
a. 5 mR
b. 14 mR
c. 17.5 mR
d. 25 mR

56. What is the radiologist's scatter exposure for the 7 minutes of fluoroscopy, if he or she stands 2 feet away as opposed to 1 foot away?
a. 35 mR
b. 17.5 mR
c. 6.25 mR
d. 2.5 mR

57. The inverse square law states that the intensity of the x-ray beam
a. varies directly with the distance.
b. varies inversely with the distance.
c. varies inversely with the square of the distance.
d. varies directly with the square of the distance.

58. If fluoroscopy exposure is suddenly stopped and a signal is heard, what would you check first?
a. kV level
b. minute timer
c. foot switch
d. door closure

59. If you are reviewing your dosimeter report, you would be reviewing your
a. medical exposure.
b. occupational exposure.
c. radon exposure.
d. background exposure.

60. The _____ interaction with matter is most common in tissue at 60 kVp.
a. Compton
b. coherent
c. pair production
d. photoelectric effect

61. What is the shortest distance from the tube to the patient when employing a C-arm unit?
a. 6 inches
b. 9 inches
c. 12 inches
d. 15 inches
e. 18 inches

62. What unit is used for the measurement of ionization in air?
a. gray
b. c/kg
c. curie
d. sievert

63. Which of the following is the safest time to radiograph a fetus?
a. third trimester
b. second trimester
c. first trimester
d. none of the above

64. Any added filtration in a radiography unit should eliminate most
a. primary radiation.
b. shorter wavelengths.
c. remnant radiation.
d. longer wavelengths.

65. Which of the following is a property of radiation that accounts for its effect on biological systems?
a. It has different energies.
b. It has no electric charge.
c. It causes ionization in matter.
d. It travels at the speed of light.

66. The degree to which recovery is possible after exposure to radiation is related to
a. total dose received.
b. degree of protraction.
c. LET of the radiation.
d. all of the above.

67. A piece of paper will provide adequate protection from which of the following types of radiation?
a. alpha
b. beta
c. gamma
d. x-rays

68. LET stands for
a. lethal effective tolerance.
b. lethal exchange table.
c. linear energy transmutation.
d. linear energy transfer.

69. Human cells are divided into two classifications, germ cells and
a. organic cells.
b. benign cells.
c. somatic cells.
d. reproductive cells.

70. What is the process of cell division of somatic cells called?
a. meiosis
b. mitosis
c. synthesis
d. multiplication

71. The direct hit theory of cell irradiation can be described by which of the following statements?
a. The DNA molecule is struck.
b. The cell nucleus is struck.
c. The cell cytoplasm is struck.
d. The cell is ionized.

72. Which of the following would describe the shape of a DNA molecule?
a. elliptical
b. hypocycloidal
c. double helix
d. circular

73. Which of the following is a type of damage that can occur to a DNA molecule when exposed to radiation?
a. change in the genetic code
b. breakage of the chromosomes
c. breakage of the DNA molecule
d. all of the above

74. What do the letters DNA stand for?
a. direct nuclear action
b. deuteron nucleus activity
c. deoxyribonucleic acid
d. deoxyribose neutral acid

75. How will oxygen retention affect the radiosenstivity of a cell?
   a. Radiosensitivity will increase.
   b. Radiosensitivity will decrease.
   c. Radiosensitivity will be eliminated.
   d. Radiosensitivity will not be affected.

76. Of the following, which tissue type is known to absorb the most radiation?
   a. air sacs
   b. bone
   c. muscle
   d. soft tissue

77. Irradiation of which of the following anatomical areas will affect the production of white blood cells?
   a. lungs
   b. liver
   c. spleen
   d. bone marrow

78. What does the acronym RBE stand for?
   a. relative biologic effectiveness
   b. radiosensitive biologic effect
   c. radioactive biological energy
   d. radiation bypass effect

79. Which of the following demonstrates that distance is one of the three cardinal principles of radiation protection?
   a. Law of Bergonie and Tribondeau
   b. inverse square law
   c. radiation power formula
   d. grid conversion formula

80. Which term describes the separation of water into hydrogen and oxygen due to irradiation?
   a. radioactivation
   b. radiolysis
   c. interaction
   d. toxicity

81. Which of the following types of patients are most radiosensitive?
   a. fetal
   b. pediatric
   c. adult
   d. geriatric

82. What is meant by the indirect effect of ionizing radiation?
   a. A genetic effect is produced.
   b. Ionization in one location may produce an effect in another.
   c. An organism other than the one irradiated may be affected.
   d. Potentially lethal metabolic dysfunction is produced.

83. Why is free radical formation in the human body so dangerous?
   a. Scatter radiation is produced in the body.
   b. All radiation shielding is penetrated.
   c. Carcinogenic tendencies have been proven.
   d. Toxic effects have been observed.

84. Which of the following would NOT affect patient dose?
   a. focal spot size
   b. mAs
   c. kVp
   d. filtration

85. Which of the following may be the result of red bone marrow damage due to radiation?
   1. Skin erythema
   2. Spinal cord myelitis
   3. Anemia
   a. 1 only
   b. 2 only
   c. 3 only
   d. 1, 2, and 3

86. What is meant by the term *interphase death* in radiobiology?
    a. Cells die before going into interphase.
    b. Cells die before leaving interphase.
    c. Cells die in between mitotic phases.
    d. The organism dies before the average cell leaves interphase.

87. What term describes cell damage from radiation that does not kill the cell?
    a. sublethal death
    b. cell repair
    c. cell healing
    d. cognition

Threshold Relationships

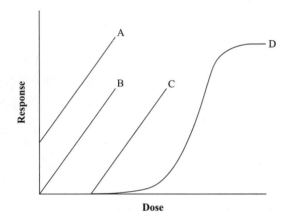

88. In reference to the preceding diagram, which line demonstrates a nonlinear dose?
    a. A
    b. B
    c. C
    d. D

89. Regarding fluoroscopy, the recommended minimum amount of filtration is
    a. 30 mm Al equivalent.
    b. 2 mm Al equivalent.
    c. 3 mm Al equivalent.
    d. 2.5 mm Al equivalent.

90. Which of the following is the active component in TLD devices?
    a. barium platinocyanide
    b. sodium lead sulfide
    c. lithium fluoride
    d. none of the above

91. What might a technologist choose to do if he or she wished to reduce the exposure of a 70 kVp, 40 mAs technique?
    a. decrease kVp by 15%
    b. decrease mAs by 30%
    c. increase kVp 15%, and halve the mAs
    d. increase mAs by 30%, and decrease kVp by 15%

92. Placing the intensification tower close to the patient does what?
    a. It increases scatter to the patient.
    b. It makes the image sharper.
    c. It reduces scatter to patient areas such as head/neck.
    d. none of the above.

93. Which of the following will improve or lower when employing beam-limiting devices?
    1. image quality
    2. patient dose
    3. production of scatter radiation
    a. 1 only
    b. 2 and 3 only
    c. 1, 2, and 3
    d. none of the above

94. Adding or increasing filtration will primarily protect which of the following?
    a. radiographer
    b. reproductive organs
    c. patient's skin
    d. patient's bone marrow

95. What should you do if preparing to shield a patient for fluoroscopy?
   a. use a shadow shield
   b. place the shield below the patient
   c. place the shield on top of the patient
   d. none of the above

96. All of the following are major steps in the acute radiation syndrome EXCEPT
   a. latent period.
   b. manifest illness.
   c. prodromal stage.
   d. dormant stage.

97. All of the following are effects from early (acute) somatic syndrome EXCEPT
   a. erythema.
   b. cancer.
   c. epilation.
   d. desquamation.

98. Inner shell ejection takes place during which interaction with matter?
   a. Compton
   b. photoelectric
   c. coherent
   d. pair production

99. All of the following use pure lead as the makeup of their barriers EXCEPT
   a. eyeglasses.
   b. aprons.
   c. gloves.
   d. protective curtain.

100. Which of the following is a self-reading dosimeter that can also provide an instantaneous measurement?
   a. TLD
   b. film badge
   c. pocket dosimeter
   d. ring badge

Chapter 5 Answers

1. **c.** Any effect in the offspring caused by radiation is a genetic mutation.
2. **a.** The prefix *hemo-* means blood, or blood-related.
3. **d.** Cataracts are well-known to occur with long-term exposure to the eye.
4. **d.** Threshold dose is correct; all other choices are ruled out by definition.
5. **c.** Leukemia is the correct answer.
6. **a.** Blood changes occur in doses of 20–25 cGy.
7. **b.** The dead man switch is well-known as an exposure switch.
8. **d.** This is the only choice that has nothing to do with radiation protection.
9. **a.** HVL is correct; it stands for half value layer.
10. **d.** Intensity has nothing to do with barrier thickness.
11. **c.** Ten or more inches is acceptable.
12. **d.** All choices are reasons for repeating a radiograph.
13. **c.** Leakage radiation is an issue for the radiologic technologist.
14. **c.** The grid is the only one located between the patient and tube.
15. **d.** A shadow shield is suspended between the patient and tube.
16. **d.** High speed screens can reduce radiation exposure.
17. **b.** The lowest ratio grid requires the lowest patient dose.
18. **b.** Seven feet is the tallest height required.
19. **b.** The protective curtain hangs from the fluoroscopy unit between the patient and fluoroscopist.
20. **a.** TVL stands for tenth value layer.
21. **b.** Rad or sievert express dose equivalents and measurements.
22. **c.** Approximately 50% is correct.
23. **b.** Time must be decreased, in an amount inversely proportional to mA.
24. **b.** Exposure is inversely proportional to the square of the distance.
25. **c.** Scatter is created any time x-rays pass through matter.
26. **b.** GSD cannot exceed 1 rem.
27. **d.** GSD cannot exceed 5 rem.
28. **a.** Distance is the primary factor.
29. **c.** Redness of the skin is skin erythema.
30. **a.** Occupational dose is measured in rem, which stands for roentgen emission man.
31. **b.** As wavelength increases, energy decreases.
32. **b.** Filters do this by removing weak, longer wavelength photons.
33. **a.** Low mAs, high kVp techniques do this.
34. **b.** The more radiation, the quicker the greatest effect is created.
35. **d.** Somatic effects refer to the exposed person.
36. **b.** Genetically significant dose is the definition of GSD.
37. **d.** Irradiating the bone marrow affects WBC (white blood cells).
38. **a.** Most scatter is formed from the matter with which it interacts.
39. **b.** Rapidly dividing cells go with the immature radiosensitivity of the cell.
40. **b.** Low WBC is a quick response to radiation.
41. **a.** Collimating is the best way to lower exposure to matter.
42. **b.** According to the Law of Bergonie and Tribondeau, young immature cells, stem cells, and highly mitotic cells are particularly radiosensitive.
43. **c.** Filters allow radiation energies to be identified.
44. **a.** TLD, as in thermo luminescent dosimeter, is correct; *thermo* indicates heat.
45. **c.** The TLD is able to record dosages as low as 5 mRem.
46. **d.** High phase, high kVp is consistent with this.
47. **b.** Sievert is a measurement of dose equivalent.

48. **c.** The lowest grid will provide this.
49. **b.** The size of a cell does not influence radiation effect.
50. **a.** One-dose delivery causes the most profound effect.
51. **a.** High kVp can allow lower mAs dosages.
52. **c.** The size of field prevalently affects the radiation amounts received.
53. **c.** Compton best answers this description.
54. **c.** The thickness is 2.5 mm aluminum equivalent.
55. **d.** 5 minutes times 5 mR/min is the entrance exposure.
56. **c.** 6.25 mR is one-fourth of 25 R (inversely proportional to square).
57. **c.** The intensity of the x-ray beam is inversely proportionate to the square.
58. **b.** The minute timer sets specific time sessions in fluoroscopy.
59. **b.** Film badges monitor occupational exposure.
60. **b.** Coherent scatter occurs below 100 kVp.
61. **c.** Twelve inches is the minimum distance.
62. **b.** Coulomb per kilogram measures ionization in air.
63. **a.** The later the better, as cells would be more mature.
64. **d.** Filtration removes weaker (longer wavelength) photons.
65. **c.** Ionization is the effect, biologically speaking.
66. **d.** All of the above matter.
67. **a.** Alpha has high LET, but lacks penetrability.
68. **d.** LET stands for linear energy transfer.
69. **c.** There are two main classes: somatic and germ cells.
70. **b.** Mitosis division is related to somatic cells.
71. **a.** The DNA molecule is struck in the direct hit theory.
72. **c.** A double helix is a DNA molecule.
73. **d.** All of the above can occur.
74. **c.** DNA stands for deoxyribonucleic acid.
75. **a.** Oxygen enhances radiosensitivity.
76. **b.** Bone absorbs radiation easiest.

77. **d.** Irradiation of bone marrow affects WBC.
78. **a.** RBE stands for relative biologic effectiveness.
79. **b.** Inverse square law demonstrates that distance is one of the three cardinal principles of radiation protection.
80. **b.** Radiolysis is water/hydrogen/oxygen separation.
81. **a.** The younger a person is, the more radiosensitive he or she is.
82. **b.** An indirect effect happens in another area.
83. **d.** Toxicity is related to free radical formation.
84. **a.** Dose is not affected by focal spot size.
85. **c.** Red bone marrow is correlated to anemia.
86. **a.** Interphase death is apoptosis.
87. **a.** Sublethal death does not kill the cell.
88. **d.** The curved line means nonlinear.
89. **d.** 2.5mm Al equivalent is recommended for fluoroscopy.
90. **c.** TLD's active component is lithium fluoride.
91. **c.** The 15% rule of radiography equals 20 mAs, 79 kVp.
92. **a.** The correct distance reduces exposure to the patient.
93. **c.** Beam-limiting devices improve all three mentioned.
94. **c.** Patient's skin exposure is improved with filtration.
95. **b.** Most fluoroscopy unit tubes are in the table, shooting upward.
96. **d.** The dormant stage is not among the major steps.
97. **b.** Cancer is a late effect.
98. **b.** Inner shell ejection is consistent with the photoelectric effect.
99. **a.** Eyeglasses are made up of lead-tinted glass.
100. **c.** A pocket dosimeter gives an immediate reading.

EQUIPMENT OPERATION AND QUALITY CONTROL

CHAPTER SUMMARY

On the ARRT exam, you will find 24 questions dedicated to equipment operation and quality control. It is necessary for radiologic technologists to have a basic understanding of these topics for a variety of reasons. They help the technologist to understand the underlying factors that contribute to image quality, as well as to understand and apply the cardinal principles of radiation protection (time, distance, and shielding), which should be reiterated across all aspects of patient care. Despite the fact that there are other sections of the ARRT exam that are greater in content, a good understanding of Equipment Operation and Quality Control is essential for proficiency in this discipline. This section will help provide a concise review of this information.

Key Terms

automatic exposure control	on/off switch
control console (operating console)	primary radiation
current	scatter radiation
effective atomic number	space charge
exposure switch	space charge effect
filament current	thermionic emission
kilovoltage control	timer
length of exposure	tissue mass density
milliamperage control	tube current

X-Ray Production

Source of free electrons: The tungsten filaments of the x-ray tube serve as the source of free electrons. Thermionic emission occurs when the filament coil is heated to a level of incandescence. The amount of heat that controls the amount of electrons emitted is controlled by the mA selected. It requires approximately 3–5 amps and 10–12 volts from the filament circuit.

Electron acceleration: Electrons are accelerated toward the target anode by the increase of potential difference between the anode and cathode. The cathode has a negative charge, and the anode is positively charged.

Focusing of electrons: The focusing cup of the cathode provides the ability to focus the electrons. The focusing cup, a negatively charged device, keeps the nonaccelerated electrons (called a space charge) close together and "corralled" prior to their being accelerated toward the anode target. The focusing cup, located on the cathode side of the tube, is constructed to surround the filament.

Deceleration of the electrons: When high-speed electrons are slowed or stopped, x-radiation is created. Most of the high-speed electrons strike the anode, which converts those electrons into x-radiation. And most of the x-radiation produced leaves the tube through the tube window toward the patient. Every time high-speed electrons bombard the target anode, a number of interactions occur. Not all of the interactions contribute to the production of the image. During this process, only 1% of the energy involved is converted to x-rays, while 99% of the energy is converted to heat. The only items needed to produce radiation are: 1) a source of electrons, 2) a way to accelerate those electrons, and 3) a way to decelerate, or slow down, the electrons.

Definitions

Filament current: the current needed to heat the filament, measured in amperage

Thermionic emission: The boiling off of electrons by heating the filament.

Space charge: The electron cloud near the filament.

Space charge effect: The repelling of electrons near the filament.

Tube current: The current needed to produce the electrons, which are needed to produce x-ray photons; tube current is measured in milliamperage. Remember mAs is the quantity of radiation and exposure, and kVp is the quality of the beam.

Properties and Characteristics of X-Rays

X-rays have the following properties:

- Highly penetrating, invisible rays
- Heterogenous or polyenergetic
- Travel in straight lines
- Travel at the speed of light
- Ionize gases
- Cause certain materials to become fluorescent
- Produce chemical and/or biological changes
- Effect photographic film
- Produce secondary and scatter radiation
- Cannot be focused by a lens
- Liberate minute amounts of heat
- Electrically neutral

Heat Production

The production of heat when producing radiation in the form of x-ray exposures is expressed by heat units (HU), which are calculated by:

$$mAs \times TIME \times kVp \times rectification\ constant\ (RC)$$

Or:

$$mAs \times kVP \times RC$$

The following are factors that must be added when calculating heat units per exposure:

Single phase = 1

Triple phase = 1.35

High Frequency = 1.40

Example: How many HUs are generated by an exposure of 70 kVp, 300 mA, and 0.15 sec on a triple phase unit?

$$70 \times 300 \times 0.15 \times 1.35 = 4,253\ HU$$

Electron-Target Interactions

Bremsstrahlung radiation (brems): *Bremsstrahlung* is a German word that means braking radiation. Most x-rays are produced from the interaction of high-speed electrons within the nuclei of target atoms of the anode.

Characteristic radiation: Along with brems, this is the other "in the tube" interaction. This interaction does not occur unless at the 70 keV range. The interaction occurs at the inner shell electron level of the target atoms. When an incoming electron collides with an inner shell electron, it dislodges or knocks the electron out of its shell. This causes a vacancy in the spot where the electron was, and this vacancy causes the atom to be in an unstable state (called an ion). This causes something like a chain reaction, where an electron from the most outer shell next to the reaction slips down to fill the initial vacancy and then the vacancy is in the outermost shell. When that outermost shell collects an atom, the atom becomes stable again. Now each time an electron slides to an inner shell to fill a vacancy, an amount of energy in the form of radiation is created. Characteristic radiation is the cumulative amount of radiation created here.

Characteristic Radiation

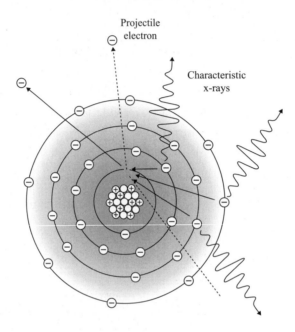

Frequency and Wavelength
Electromagnetic radiation is radiation that has both electric and magnetic properties. Frequency is the number of cycles per second in an electromagnetic wave. A wavelength is that distance between any two cycles in a wave, and velocity is the speed, which is measured in the frequency multiplied by the wavelength:

$$\text{Velocity} = \text{Frequency} \times \text{Wavelength}$$

Photon Interaction with Matter
Compton effect: These interactions account for most of the scatter radiation in an exposure in the diagnostic range. They also account for fog on a radiograph. A Compton occurs when a photon of less energy interacts with an electron in an outer shell of an atom.

In this case, an x-ray interacts with an outer shell electron (1). The electron (2) is ejected (now known as the Compton electron). The now scattered x-ray (3) will continue in a different direction and may have other interactions, ionizing other atoms.

The Compton Effect

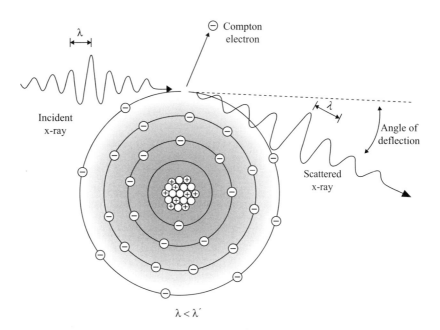

$\lambda < \lambda'$

Photoelectric effect: This effect occurs between a photon and an inner shell electron. In this interaction, the photon is completely absorbed by the atom, and an electron is ejected from the atom with the energy of the original photon minus the binding energy. This interaction occurs from low energy ranges, usually below the 70 kV range.

The Photoelectric Effect

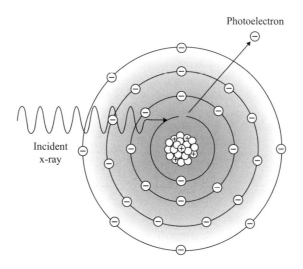

Coherent scatter: Also called Rayleigh or classical scatter, this interaction happens whenever photons of low energy scatter from an atom without any energy loss, hence the name *coherent*. This is not a major occurrence, as it happens less than 5% of all of the photon interactions. In addition, there is no change in the construction of the targeted atom. The energy of the ejected photon is the same as the energy of the incident photon.

Coherent Scatter

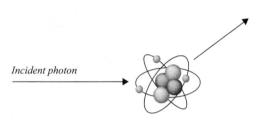

Incident photon

Pair production: This interaction is typically used in discussions and is often a choice in questions regarding the interactions of energy with matter. A pair production interaction requires an energy level of at least 1.02 MeV (million electron volts), and since this level is only possible at the therapeutic level (techniques in radiation therapy), it is more than unlikely to occur at the diagnostic level.

X-Ray Tube and Beam Characteristics

Parts of the Tube

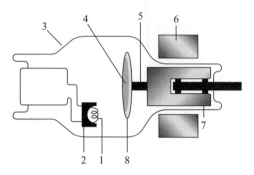

Parts of the Tube

(1) Cathode/filament: Negative terminal of tube; supplies the electrons when heated.

(2) Focusing cup: Reduces the spreading of the space charge.

(3) Glass tube (Pyrex): Holds the cathode and anode; keeps vacuum intact.

(4) Rotating anode (Disc): Attracts the electrons; produces x-rays; is positively charged.

(5) Anode stem: Made of molybdenum.

(6) Stator: Located outside the glass tube; part of the induction motor; series of electromagnets equally spaced around the neck of the tube.

(7) Rotor: A copper shaft inside the glass tube that spins when induction occurs through interaction of the stator and the external stator.

(8) Target of anode: The area of the anode face that the electrons strike during exposure. It should be noted that tungsten is the most popular material that makes up the target material. Any target material needs to have a high atomic number, thermal conductivity, and a high melting point. Along with tungsten, other popular materials used in the construction of an x-ray tube are: molybdenum, copper, and graphite.

Definitions

Remnant photon: All of the x-ray photons that reach their destination (the film) after passing through the object being radiographed; the photons that have contributed to the image.

Scatter radiation: A process in which the x-ray photons undergo a change of direction after interacting with the atoms of an object.

Attenuation: Any process that decreases the intensity of the beam that was directed toward some destination.

Attenuation

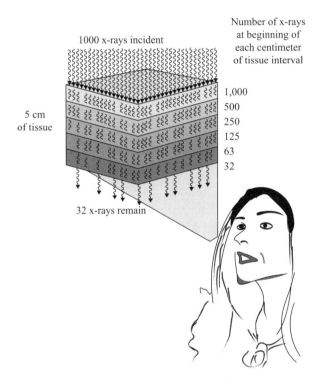

1000 x-rays incident

5 cm of tissue

Number of x-rays at beginning of each centimeter of tissue interval

1,000
500
250
125
63
32

32 x-rays remain

Interaction of x-rays by absorption and scatter is attenuation. Here the x-ray beam has been attenuated 97% while 3% of the x-rays have been transmitted.

Primary radiation: Any photon emerging from the tube that has not yet interacted with tissue

Tissue mass density: Different sections of body mass have equal thicknesses, yet different mass densities.

Tissue Mass Density

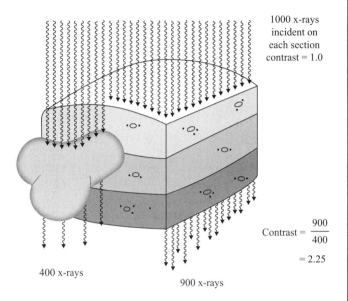

1000 x-rays incident on each section contrast = 1.0

400 x-rays

900 x-rays

$$\text{Contrast} = \frac{900}{400}$$

$$= 2.25$$

Effective atomic number: When the effective atomic number of adjacent tissues is very high, subject contrast is very high.

Effective Atomic Numbers of Materials Relevant to Diagnostic Radiology

Type of Substance	Effective Atomic Number
HUMAN TISSUE	
Fat	6.3
Soft tissue	7.4
Lung	7.4
Bone	13.8
CONTRAST MATERIAL	
Air	7.6
Iodine	z53
Barium	56
OTHER	
Concrete	17
Molybdenum	42
Tugsten	74
Lead	82

Control console (operating console): The control console controls the voltage, the "push," which causes electricity to flow. It is also called electromotive force (EMF).

X-Ray Unit Operating Console

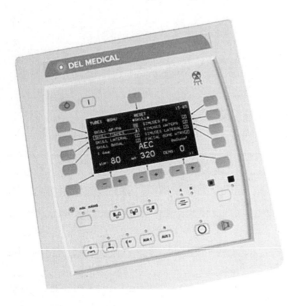

From here the operator controls the voltage, which causes electricity to flow.

Current: The amount of electricity per second through the tube.

Length of exposure: The duration of the exposure in seconds.

On/off switch: A push button on the control console and one or more toggle switches on the wall (called the wall breaker).

Milliamperage control: Controls current; the unit is the milliampere (abbreviated mA). Range of current is approximately 25 to 1,000 mA.

Exposure switch: A two-part device (usually two separate push buttons). The first function is the ready function, which prepares the tube for producing x-rays. After a short delay, pressing the exposure switch begins the production of x-rays. You must keep the exposure switch depressed until the timer ends the exposure!

Kilovoltage control: Controls voltage; the unit is the kilovolt (abbreviated kV). Range of voltage is 50 to 150 on general-purpose diagnostic equipment.

Timer: The manual timer controls the length of the exposure, measured in seconds (abbreviated s).

Automatic exposure control: An automatic timer that detects the amount of radiation passing through the patient and terminates the exposure after the appropriate time (abbreviated AEC; often called phototiming)

X-Ray Imaging Circuitry

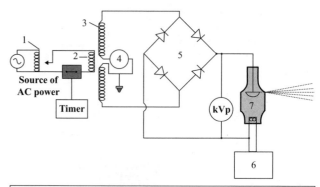

X-ray imaging circuitry.
(1) Autotransformer
(2) Primary side of the transformer
(3) Secondary side of the transformer
(4) mA meter
(5) Full-wave rectifiers
(6) Focal spot selector
(7) X-ray tube

Voltage Waveforms

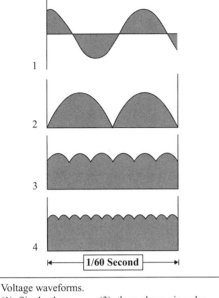

Voltage waveforms.	
(1) Single phase	(3) three phase, six pulse
(2) Full wave	(4) three phase, twelve pulse

Automatic Exposure Control

The main purpose of AEC is to eliminate the need for the radiographer to set an exposure time. mA and kVp must still be set. To terminate the radiation exposure time, AEC devices provide diagnostic-quality exposures only for structures positioned directly above the ionization chambers. AEC is used when performing abdominal radiography and most other radiographic exams.

The most popular AEC device is three chambers with the middle chamber at the center of the image receptor and the right and left chambers slightly higher, as depicted in the following figure.

Three-Chamber AEC Device

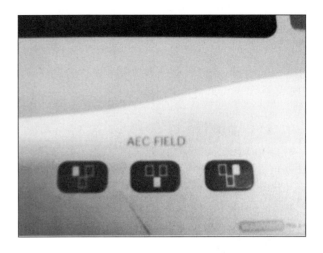

This setup places the center chamber below the doudenum and transverse colon for most abdominal examinations, eliminating the issue with bowel gas being placed over the chamber, and thus causing the exposure to terminate too soon. For chest radiography, the two outside cells should be used. This will help to cause termination of exposure after the two lungs are sufficiently exposed, as opposed to termination of exposure when the spine or central area is sufficiently exposed.

There are seven different combinations with the three-chamber systems. Experience in using AEC units will help the radiographer to perfect his or her skills and selections in this area.

Density controls: These should be used to compensate for patient part thickness or kVp changes. This is the −1, −2, N, +1, and +2 settings.

Averaging: When exposure is occurring, the signals from the cells are sent to a special operational amplifier, which sums the voltages received from each activated cell and divides them by the number of activated cells. When more than one cell is activated, the cell receiving the most radiation will contribute the greatest electrical signal and have the most influential impact on the overall exposure.

Minimum reaction (response) time: Minimum reaction time is determined by the length of time necessary for the AEC to respond to the radiation and for the generator to terminate the exposure. Backup time should be set at 150% of the anticipated manual exposure mAs.

Fluoroscopic Unit

Fluoroscopy involves dynamic viewing, which means viewing as it actually happens, or in real time. The continuous beam lets you see moving images.

Generally, fluoro is imaged at 5 mA, but patient dose is still not reduced because the imaging time is much longer, usually in minutes. The kVp level during fluoroscopy depends mainly on the thickness of the body part being imaged. The equipment allows the fluoroscopist to pick a brightness level that is acceptable. This is usually selected automatically by fluctuating the mAs and kVp, or both. This feature is referred to as automatic brightness control (ABC).

Image Intensification Tube

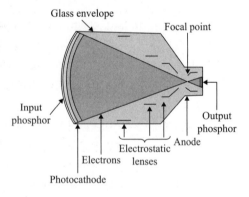

Remnant photons transmit through the glass envelope and interact with the input phosphor (cesium iodide). When this happens, its energy is converted into visible light. The light then interacts with the photocathode. What does this do?

Photo =light, and *cathode* =electron emission. It takes many light photons to cause the emission of one electron. The number of electrons emitted by the photocathode is directly proportional to the intensity of the light reaching it. So, this all means that the number of electrons emitted is proportional to the intensity of incident image-forming x-ray beams.

The anode is a circular plate with a hole in the middle to allow the electrons through to the output phosphor made of zinc cadmium sulfide. This is where the electrons interact and produce light.

Television Camera Tube

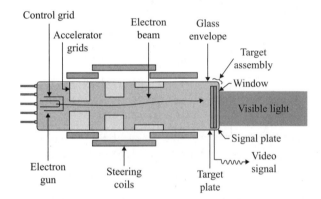

How the television camera tube works: The electron gun (cathode) is a heated filament that supplies the constant electron current by thermionic emission. The control grid forms the electrons into an electron beam and helps to accelerate the electrons to the anode. The electrostatic grids also help to accelerate the electrons to the anode. The steering coils control the size of the electron beam and its position.

Target assembly: The target assembly consists of three layers sandwiched together. From outside to inside they are:

1. Face plate (window): the thin part of the glass envelope

2. Signal plate: coated on the inside of the window, a thin layer of metal or window. This layer is thin enough to transmit light, but thick enough to conduct electricity (metal or graphite).

3. Target or photoconductive layer: This layer is made of antimony trisulfide (vidicon) and is photoconductive (which means that when it is illuminated, it conducts electrons, and it behaves as an insulator when dark).

When light from the output phosphor of the II tube strikes the window, it is transmitted through the signal plate to the target. Only video signal is produced where the electron beam and light are hitting the target. If the area of the target is dark, there is no video signal.

Vignetting: A reduction of brightness at the periphery of the image.

Spatial resolution: Refers to the ability to image small objects that have high subject contrast; for example, bone/soft tissue combination.

Contrast resolution: The ability to distinguish anatomic structure of similar subject contrast.

Television picture tube (TPT): In other words, the TV monitor (CRT).

CRT (Cathode Ray Tube, or TV Monitor)

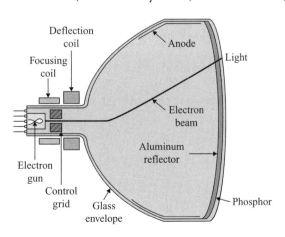

The video signal is amplified and transmitted by cable to the television monitor, where it is transformed back into a visible image.

The TPT takes its information directly from the television camera tube and converts the information back to an adequately sized image for the fluoroscopist to examine. In addition, the brightness and contrast can be manipulated for proper real-time viewing. The two controls on the TV monitor are for brightness and contrast. There is no audio or channel selection.

Portable Light-Duty and Full-Power Institutional Mobile Units

All portable machines should have an exposure cord 6–12 feet long. All portable light-duty and some full-power machines get their power from wall outlets: you have to plug them in. Most units can use 110 volts, but bigger units require 220 volts. For units drawing power from outlets, the voltage compensator must be accurately adjusted immediately prior to exposure. There are two types of portable/mobile units: battery-powered units and capacitor discharge units.

Capacitor discharge units are devices which are plugged in and absorb a certain amount of voltage that is used for one exposure. A capacitor is an electrical device that stores a charge, or electrons. The standard 110 or 220 volts of power are fed into the transformer in the unit. The unit must be plugged in and immediately store a charge before each exposure. Once the capacitor is charged, it can be discharged through the x-ray tube. These units are small and easy to move about. It is not practical to charge the unit in the department, and then take it to another destination for an exposure, because the charge rapidly leaks away from the capacitor.

In battery-powered units, a standard power supply is used to charge large-capacity nickel-cadmium batteries. These units are always larger and heavier due to the battery. Once charged, they can take many exposures from patient to patient and room to room. There is a motor to operate the movement of the heavy units, and they are usually equipped with a dead man switch on the motor.

Battery-Powered Mobile Unit

Capacitor Discharge Mobile Unit

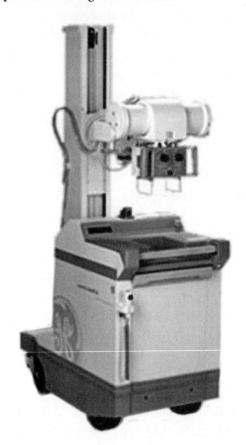

Quality Control of Radiographic Equipment

Quality control of radiographic equipment involves various types of testing. These tests cover all equipment areas, including the operating console, the tube, the x-ray table, and all radiographic accessories. These tests ensure proper working of the machines and all of their parts.

Testing kVp Voltage Accuracy

The variation between the selected kVp and the actual beam quality must be within a 5% range. Because of the importance of controlling radiographic contrast on images, there are specific tests that can be done to monitor its variances.

The various types of tests that can be done include:

- Wisconsin test cassette
- Digital kilovolt meter
- Ardan and Crooke cassette

Wisconsin Test Cassette Device

Central Ray Ali Central Ray Alignment Test

The following image is of a central ray alignment test. It is comprised of a plastic cylinder with steel balls used for testing the central beam angle. The two $\frac{1}{16}$ inch steel balls are located directly above and below each other and are imbedded within the 6-inch plastic cylinder. The cylinder is placed on top of the beam template, which is placed on top of a cassette. If the table and tube are aligned correctly, you should only see one ball in the center of the template because the two balls would be superimposed perfectly on top of each other.

Timer Accuracy

This is an important test to determine if the quantity of radiation is accurate (mAs). This test is only relevant when testing single phase units, as single phase is emitted in a pulsating action. The allowable variation on timer accuracy is a 5% variability on exposure times greater than 10 milliseconds and 20% variability on exposure times less than 10 milliseconds. A spin top test is one of the older methods used. The only device needed is a simple two-piece unit in which a toplike device with a single hole is spun manually on an axis. Depending on which time stop is being tested, there will be a number of dots that will appear on the radiograph, which coincides with the exposure time. For example, if you are testing the 0.10 second time, and the machine is a half-wave rectified unit, the correct number of dots you should see is six ($0.10 \times 60 = 6$ dots). This is depicted in the following image.

Spin Top Test

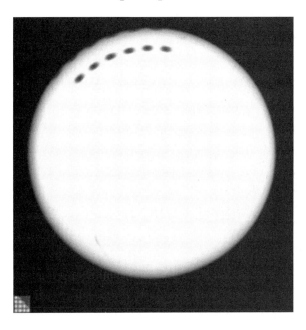

This example demonstrates six dots from a 0.10 second exposure.

High-Contrast Resolution in Fluoroscopy

The square mesh device with the triangular screenlike sections in the following image is used to test small, thin, black and white areas, which is how high-contrast resolution is measured. With the kVp setting at its lowest level, the mesh device is taped to the bottom of the image intensifier. When this device projects an image, you can compare the number of holes that can be seen on the periphery of the image to how many can be seen in the center of the image. The lead numbers within each of the triangular areas represent the holes per inch.

Wire Mesh Test

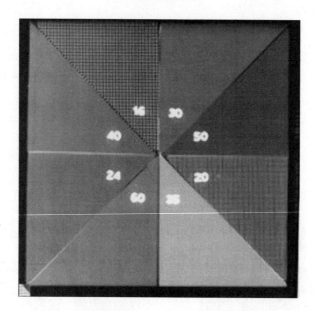

Screen-Film Contact

A simple copper wire device—that looks like a piece of screen, except the wiring is thicker and is made of copper wiring—is used to demonstrate screen-film contact in cassettes. To utilize, you simply place the wire mesh device on top of a film-loaded cassette and radiograph it. If the cassette has optimal film-screen contact, you will observe a consistent pattern throughout the radiograph produced, as depicted in Example A. An inconsistent, blurry pattern, demonstrating poor film-screen contact, is demonstrated in Example B.

Example A: **Optimal Contact**   *Example B:* **Poor Contact**

Spatial Resolution

Spatial resolution is the ability to separate different levels so that two like structures can be distinguished. A line pair test is done to demonstrate this. You simply place a device (seen in the following image) directly on a cassette and take an exposure. The device is made up of small, thin strips of lead, precisely spaced, and varying in thickness. The lead strips are placed in a specific intervals of "line pairs per mm."

Line Pair Test Device

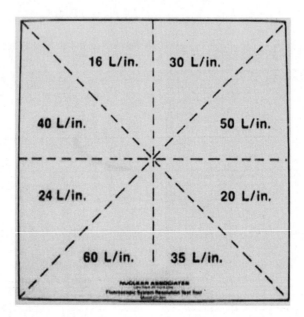

Chapter 6 Review Questions

Please use the following depiction of an image intensification tube to answer questions 1 through 4.

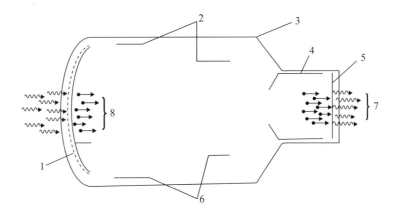

1. In the figure, #7 depicts the
   a. input phosphor.
   b. x-ray beam.
   c. light image.
   d. anode.

2. In the figure, #5 depicts the
   a. x-ray beam.
   b. light image.
   c. electron image.
   d. output phosphor.

3. In the figure, #3 depicts the
   a. electrostatic lenses.
   b. anode.
   c. output phosphor.
   d. glass envelope.

4. In the figure, #1 depicts the
   a. input phosphor.
   b. x-ray beam.
   c. light image.
   d. anode.

5. An electron gun is one component of which of the following?
   1. Television monitor
   2. Vidicon camera
   3. Image intensifier

   a. 1 and 2 only
   b. 1 and 3 only
   c. 2 and 3 only
   d. 1, 2, and 3

6. Phototiming equipment automatically
   a. increases processing time.
   b. lowers the mA when needed.
   c. ends the x-ray exposure.
   d. increases the kV.

7. Mobile techniques often differ from techniques used in the radiology department because
   a. available mA and kV are limited.
   b. SID may be longer or shorter than usual.
   c. the patient may be more severely ill.
   d. all of the above.

8. Portable radiographic equipment which does not have to be plugged in during the exposure is
a. operated by magnetism.
b. of very low capacity.
c. battery powered.
d. of the capacitor discharge type.

9. The input phosphor of the image intensifier tube converts a(n)
a. light image to an x-ray image.
b. x-ray image to a light image.
c. light image to an electron image.
d. electron image to a light image.

10. There are two reasons why the output image from the image intensifier is brighter; one is because of the acceleration of the electron image, the other is because of the
a. conversion of light to x-ray.
b. minification of the image.
c. amplification in the viewing system.
d. use of the magnification.

11. What is a typical mA for the use of fluoroscopy?
a. 100 mA
b. 20 mA
c. 5 mA
d. 0.2 mA

12. In the image intensifier, the electron image is reduced in size by the action of the
a. input phosphor.
b. video camera.
c. voltage between the anode and cathode.
d. focusing electrodes (electrostatic lenses).

13. The vidicon and plumbicon television cameras convert
a. x-ray to light.
b. electricity to light.
c. light to x-ray.
d. light to electricity.

14. In most cases, there is a relationship between screen speed and screen resolution. Which screen resolution below is probably associated with the highest screen speed?
a. 7 lp/mm
b. 10 lp/mm
c. 12 lp/mm
d. 15 lp/mm

15. Computed radiography systems, compared to film-screen systems
a. require more exposure.
b. have less latitude.
c. have better spatial resolution.
d. all of the above.

16. When a tube has a smaller target angle, it usually has a
a. higher rotation speed.
b. lower rotation speed.
c. smaller effective focal spot.
d. larger effective focal spot.

17. If quantum mottle is present on a radiographic image, it indicates that
a. too few x-ray photons were used.
b. the intensifying screens are damaged.
c. the transport rollers are dirty.
d. the patient moved slightly.

18. On the Fuji CR system, an S number that is lower than the acceptable range indicates
   a. low subject contrast.
   b. high subject contrast.
   c. insufficient exposure.
   d. excessive exposure.

19. If two structures that are far apart appear to be closer together on the radiograph, it indicates the presence of
   a. elongation.
   b. motion.
   c. magnification.
   d. spatial distortion.

20. Which test is used for film-screen contact?
   a. Wisconsin test
   b. wire mesh test
   c. star test pattern
   d. none of the above

21. What is the number of HU which is produced by a three phase, six pulse exposure of 800 mA, 0.01 sec, and 90 kV?
   a. 720
   b. 972
   c. 1,015
   d. 1,440

22. The tube anode will lose heat most rapidly when it is
   a. cool.
   b. slightly warm.
   c. very warm.
   d. very hot.

23. Which tube anode is likely to have the greatest heat capacity?
   a. 4" disc with 14E target angle
   b. 4" disc with 9E target angle
   c. 5" disc with 9E target angle
   d. 5" disc with 14E target angle

24. Which of these would be used to evaluate timer accuracy on three phase equipment?
   a. densitometer
   b. spin top test
   c. dosimeter
   d. conventional spin top test

25. A copper mesh test is used to evaluate
   a. collimator accuracy.
   b. geometric unsharpness.
   c. film-screen contact.
   d. beam quality.

26. A plastic cylinder with two small $\frac{1}{16}$ inch metal balls imbedded in it tests for
   a. collimator accuracy.
   b. geometric unsharpness.
   c. film-screen contact.
   d. beam quality.

27. Timer accuracy and reproducibility should NOT exceed
   a. + 0.1%.
   b. + 5%.
   c. + 15%.
   d. + 50%.

28. Perpendicularity of the beam is especially necessary in Bucky radiography because it prevents
  a. grid cutoff.
  b. magnification.
  c. increased penumbra.
  d. motion unsharpness.

29. Malalignment of the center of the light field with the center of the x-ray field must NOT be greater than
  a. 1% of the SID.
  b. 2% of the SID.
  c. 10% of the SID.
  d. 50% of the SID.

30. Using which of these would probably reduce quantum mottle?
  a. a higher processing temperature
  b. lower mAs and higher screen speed
  c. higher mAs and lower kVp
  d. longer SID

31. Which set of technique factors will produce the most heat in the anode of the x-ray tube? (All produce the same density.)
  a. 50 mAs, 92 kVp, single phase
  b. 100 mAs, 80 kVp, single phase
  c. 50 mAs, 84 kVp, three phase
  d. 100 mAs, 72 kVp, three phase

32. What is true of a capacitor discharge portable machine?
  a. It does not have to be plugged in.
  b. It stores enough energy for about 50 chest exams.
  c. It must be charged immediately before each exposure.
  d. It allows higher mA and kV than other types.

33. In an under-table fluoro system, which structures are under the table?
  a. image intensifier and grid
  b. image intensifier and video camera
  c. x-ray tube and collimator
  d. x-ray tube and image intensifier

34. When phototiming an AP or PA chest, you should select
  a. one side field.
  b. both side fields.
  c. all three fields.
  d. none of the above.

35. Which type of mobile x-ray machine is the most common?
  a. battery powered
  b. conventional
  c. capacitor discharge
  d. hybrid

36. Which of the following best describes the coherent interaction?
  a. inner shell electron dislodged from its orbit
  b. outer shell electron dislodges from its orbit
  c. creation of a positron and negatron
  d. incident electron totally absorbed causing atom to vibrate

37. A rotating anode works off a(n) _____ motor.
  a. indirect
  b. direct
  c. induction
  d. noninduction

38. Which interaction does not occur at the diagnostic range?
   a. pair production
   b. coherent
   c. photoelectric
   d. Compton

39. Which is not associated with the anode?
   a. target
   b. stem
   c. focusing cup
   d. stator

40. The anode stem is usually made of
   a. aluminum.
   b. tungsten.
   c. copper.
   d. molybdenum.

41. Characteristic radiation will not occur unless at the _____ keV range.
   a. 40
   b. 50
   c. 60
   d. 70

42. An ion is an atom with
   a. extra neutrons.
   b. extra electrons.
   c. extra protons.
   d. none of the above.

43. Velocity = Frequency × _____ ?
   a. current
   b. mAs
   c. kVp
   d. wavelength

44. Which occurs during characteristic radiation?
   a. heat
   b. a projectile electron
   c. a negatron
   d. a positron

45. Which of the following is NOT needed to produce radiation?
   a. anode
   b. cathode
   c. electrons
   d. heat

46. Which of the following is the quantity of exposure?
   a. mAs
   b. kVp
   c. electrons
   d. heat

47. Which of the following is NOT a property of an x-ray?
   a. affects photographic film
   b. electrically neutral
   c. electrically negative
   d. ionizes gases

48. In German, *bremsstrahlung* means
   a. heated.
   b. braking.
   c. electric.
   d. speeding.

49. When the filament coil is heated and the electrons are boiled off, this describes
   a. the focusing of electrons.
   b. thermionic emission.
   c. the acceleration of electrons.
   d. none of the above.

50. The cathode has a _____ charge, and the anode has a _____ charge.
    a. positive/negative
    b. neutral/negative
    c. negative/positive
    d. none of the above

51. The focusing cup keeps the electrons
    a. together.
    b. apart.
    c. positively charged.
    d. none of the above.

52. During an exposure, 99% of the energy created is in the form of
    a. heat.
    b. radiation.
    c. radon.
    d. amperage.

53. Photons that contribute to the image are referred to as
    a. scatter.
    b. attenuated.
    c. remnant.
    d. none of the above.

54. Which of the following is the target substance of choice for an x-ray tube?
    a. tungsten
    b. copper
    c. aluminum
    d. glass

55. Primary photons, minus the remnant photons, equal
    a. projected photons.
    b. scattered photons.
    c. absorbed photons.
    d. attenuated photons.

56. Bone has an atomic number of
    a. 6.3.
    b. 7.0.
    c. 13.8.
    d. 10.

57. Fat has an atomic number of
    a. 6.3.
    b. 7.0.
    c. 7.4.
    d. 10.

58. Which of the following is not on the operating console of an x-ray machine?
    a. focusing cup
    b. on/off button
    c. milliamperage control
    d. exposure switch

59. If only the central photocell was selected when performing a PA Chest exam, what would be the outcome?
    a. no exposure
    b. over-penetration of lung fields
    c. under-penetration of lung fields
    d. diagnostic radiograph

60. With the three-chamber system of AEC, there are _____ different combinations.
    a. 3
    b. 5
    c. 6
    d. 7

61. For phototiming, backup time should be set at _____ of the anticipated manual exposure.
    a. 50%
    b. 100%
    c. 150%
    d. 200%

62. The term *dynamic viewing* is most associated with
   a. stereoscopic radiography.
   b. fluoroscopy.
   c. tomography.
   d. portable/mobile x-ray.

63. ABC stands for
   a. anode blooming control.
   b. automatic brightness control.
   c. automatic brightness center.
   d. none of the above.

64. Which of the following is not associated with the image intensifier?
   a. output phosphor
   b. photocathode
   c. input phosphor
   d. electron gun

65. Which comes first in the image intensifier?
   a. input phosphor
   b. photocathode
   c. electrostatic lenses
   d. output phosphor

66. Zinc cadmium sulfide is the material which makes up the
   a. photocathode.
   b. input phosphor.
   c. output phosphor.
   d. anode.

67. The term *capacitor discharge* is most associated with
   a. fluoroscopy.
   b. mobile radiography.
   c. tube current.
   d. none of the above.

68. In battery-powered mobile units, the type of battery used is usually
   a. nickel-cadmium.
   b. cellular.
   c. QC type.
   d. capacitation.

69. All of the following are kVp voltage accuracy tests EXCEPT
   a. Wisconsin test.
   b. digital kilovolt meter.
   c. kilo-equivalent method.
   d. Ardan and Crooke.

70. The output phosphor of an image intensification emits what?
   a. electrons
   b. light
   c. x-rays
   d. none of the above

71. In regular x-ray techniques, mAs is generally _____ in fluoroscopy.
   a. higher than
   b. lower than
   c. the same as
   d. none of the above

72. The most popular type of exposure activation switch for fluoro use is the
   a. relay switch.
   b. circuit switch.
   c. micro switch.
   d. dead man switch.

73. Within the image intensifier, the purpose of the photocathode is
   a. to change electrons to x-rays.
   b. to change light to electrons.
   c. to change electrons to light.
   d. none of the above.

74. Minification gain is
   a. the square of the ratio of the input screen to the output screen.
   b. the product of the input screen to the output screen.
   c. the square root of the input screen to the output screen.
   d. none of the above.

75. The term most related to the action of thermionic emission is
   a. phosphorescence.
   b. incandescence.
   c. luminescence.
   d. fluorescence.

76. It is the _____ that accelerates the electrons toward the output phosphor.
   a. photocathode
   b. electron gun
   c. anode
   d. electrostatic lens

77. Tube rating charts
   a. provide you with the proper technique for patients.
   b. tell you how to operate the x-ray machine.
   c. provide you with the safe technical factors for the x-ray tube for a single exposure.
   d. none of the above.

78. The device that allows free flow of electrons is a(n)
   a. insulator.
   b. transducer.
   c. conductor.
   d. resistor.

79. The QA test that checks the size of the focal spot is
   a. Wisconsin test.
   b. SMPTE test.
   c. step wedge test.
   d. star test pattern.

80. An area of blurring as a result of a wire mesh test demonstrates that
   a. you have a warped or damaged cassette.
   b. the intensifying screens are old.
   c. the kVp meter is not working.
   d. the mAs meter is not working.

81. Eventually, there will be failure of an x-ray tube to produce, and this reason is most related to
   a. anode damage (pitting).
   b. excessive heat.
   c. filament burnout.
   d. none of the above.

82. Why is there a warm-up procedure for x-ray machines?
   a. It increases the production of the tube.
   b. It strengthens the space charge.
   c. It properly warms the anode.
   d. All of the above.

83. What is the primary purpose of phototiming?
   a. so the technologist does not need to set a kVp
   b. so the technologist does not have to position the patient
   c. so the technologist does not have to set an exposure time
   d. all of the above

84. What interacts with the input phosphor of the image intensification tube?
a. scatter photons
b. remnant photons
c. absorbed photons
d. none of the above

85. The input phosphor of the image intensifier is made of
a. zinc cadmium sulfide.
b. barium lead sulfate.
c. tungsten.
d. cesium iodide.

86. The adjustment to vision in dim light is known as
a. duskation.
b. dark adaptation.
c. rodation.
d. none of the above.

87. All of the following are parts of the television camera tube EXCEPT
a. target plate.
b. electron gun.
c. steering coils.
d. photocathode.

88. The gas found in ion chambers of AEC devices is
a. nitrogen.
b. oxygen.
c. Freon.
d. free air.

89. What would be the primary concern of a technologist who is overusing an x-ray machine?
a. allowing the machine to cool
b. being exhausted from working so hard
c. inconsistent kVp because of overuse
d. all of the above

90. How many heat units are produced for three exposures with a 10 mAs, 80 kVp technique?
a. 2,400 HU
b. 1,200 HU
c. 800 HU
d. 80 HU

91. Which of the following has the highest atomic number?
a. water
b. bone
c. muscle
d. soft tissue

92. What substance has an atomic number of 82?
a. concrete
b. molybdenum
c. tungsten
d. lead

93. Generally speaking, what does the heating of the filament regulate?
a. kVp
b. focal spot size
c. mAs
d. none of the above

94. What is the numerical factor that needs to be added to the product to calculate heat units for a three phase, six pulse unit?
a. 11.35
b. 1.40
c. 1.41
d. 1.35

95. What alloy is commonly used in combination with tungsten in the construction of an anode?
a. copper
b. tin
c. aluminum
d. rhenium

96. Which of the following best describes the wavelength of an x-ray beam?
a. homochromatic
b. monoplagic
c. heterogenous
d. homogenous

97. Which interaction will affect contrast the most?
a. Rayleigh
b. photoelectric
c. pair production
d. Compton

98. Following a Compton interaction, _____ radiation is usually produced.
a. brems
b. characteristic
c. scatter
d. radon

99. With respect to magnification mode, the field of view is _____ in the smallest setting.
a. the same
b. larger
c. smaller
d. none of the above

100. Generally speaking, implementing the 15% rule will _____ heat units on an exposure.
a. increase
b. decrease
c. not affect
d. none of the above

Chapter 6 Answers

1. **c.** 7 demonstrates the light image.
2. **d.** 5 demonstrates the output phosphor.
3. **d.** 3 represents the glass envelope.
4. **a.** 1 demonstrates the input phosphor.
5. **a.** An electron gun is a part in television monitors and vidicon cameras.
6. **c.** AEC automatically terminates the exposure.
7. **d.** All of the above are differences.
8. **c.** Battery-powered units do not have to be plugged in for an exposure.
9. **b.** Input phosphor converts x-rays to light.
10. **b.** Another reason is the minification of the image.
11. **c.** 5 mA is a typical selection for fluoroscopy.
12. **d.** Focusing electrodes reduce the image size.
13. **d.** TCTs convert light to electricity.
14. **a.** The lowest screen resolution is associated with the highest screen speed.
15. **a.** Computer systems need more exposure.
16. **c.** Smaller target angles result in smaller effective focal spots.
17. **a.** Quantum mottle equals too few photons.
18. **d.** When the S number too low, it indicates excessive exposure.
19. **d.** When two structures that are far apart appear to be closer together on the radiograph, it indicates spatial distortion.
20. **b.** The wire mesh test is used to test film-screen contact.
21. **b.** 972 heat units.
22. **d.** The hotter it is, the quicker it will cool.
23. **c.** The larger the disc, the smaller the tube angle.
24. **b.** The spin top test tests timer accuracy on three phase equipment.
25. **c.** A copper mesh test is used to test screen contact.
26. **a.** The item described tests for collimator accuracy.
27. **b.** Timer accuracy and reproducibility should not exceed +5%.
28. **a.** Perpendicularity is necessary to avoid reciprocating the grid in the table.
29. **b.** Malalignment should not be greater than 2% of the source-to-image distance.
30. **c.** Using higher mAs and lower kVp reduces quantum mottle.
31. **d.** Higher mAs, kVp, and three phase over single phase will produce the most heat in the anode of the x-ray tube.
32. **c.** The capacitor discharge portable machine must be charged before each exposure.
33. **c.** The tube and collimator are under the table.
34. **b.** Both sides should be selected when phototiming an AP or PA chest.
35. **a.** Battery powered is the most common.
36. **d.** Describes the coherent effect
37. **c.** A rotating anode works off of an induction motor.
38. **a.** Pair production is only in the therapeutic range.
39. **c.** The focusing cup is not associated with the anode.
40. **d.** The anode stem is usually made of molybdenum.
41. **d.** Characteristic radiation will not occur unless at the 70 keV range.
42. **b.** An ion atom contains extra electrons.
43. **d.** Wavelength is correct.
44. **b.** A projectile electron occurs during characteristic radiation.
45. **d.** Heat is not needed to produce radiation.
46. **a.** The quantity of exposure is expressed in mAs.
47. **c.** An x-ray is not electrically negative.
48. **b.** Braking is the correct answer.
49. **b.** Thermionic emission takes place when the filament coil is heated and electrons are boiled off.
50. **c.** The cathode has a negative charge, the anode has a positive charge.
51. **a.** The focusing cup keeps electrons together.
52. **a.** Ninety-nine percent of the energy created is heat.

53. **c.** Remnant photons contribute to the image.
54. **a.** Tungsten is the substance of choice.
55. **d.** Attenuated photons are primary photons minus the remnant photons.
56. **c.** 13.8 is the atomic number for bone.
57. **a.** 6.3 is the atomic number for fat.
58. **a.** The focusing cup is not on the console.
59. **b.** Central photocell will not terminate the exposure until the thoracic spine is properly penetrated thus overexposing the lung fields.
60. **d.** There are seven combinations with the three-chamber system of AEC.
61. **c.** Backup time should be set to 150%.
62. **b.** Fluoroscopy is most associated with dynamic viewing.
63. **b.** ABC stands for automatic brightness control.
64. **d.** The electron gun is not associated with the image intensifier.
65. **a.** Input phosphor comes first in the image intensifier.
66. **c.** Output phosphor is made up of zinc cadmium sulfide.
67. **b.** Capacitor discharge is a type of mobile radiography unit.
68. **a.** Nickel-cadmium is the type of battery most used.
69. **c.** Kilo-equivalent method is not a kVp voltage accuracy test.
70. **b.** The output phosphor of an image intensification emits light.
71. **b.** mAs is generally lower with regular x-ray techniques.
72. **d.** The dead man switch is the most popular type of exposure activation switch for fluoro use.
73. **b.** In the image intensifier, the photocathode changes light to electrons.
74. **a.** It's the square of the ratio of the input screen to the output screen.
75. **b.** Incandescence is most related to the action of thermionic emission.
76. **d.** The electrostatic lens accelerates the electrons toward the output phosphor.

77. **c.** Tube rating charts provide you with the safe technical factors for the x-ray tube for a single exposure.
78. **c.** The conductor allows the free flow of electrons.
79. **d.** The star test pattern checks the size of the focal spot.
80. **a.** A warped or damaged cassette creates an area of blurring.
81. **c.** Filament burnout causes failure of an x-ray tube to produce.
82. **c.** Warming up the x-ray machines properly warms the anode.
83. **c.** The technologist does not have to set an exposure time when using phototiming.
84. **b.** Remnant photons interact with the input phosphor of the image intensification tube.
85. **d.** The input phosphor of the image intensifier is made of cesium iodide.
86. **b.** Dark adaptation is the adjustment to vision in dim light.
87. **d.** A photocathode is not a part of the television camera tube.
88. **d.** Free air is found in ion chambers of AEC devices.
89. **a.** The technologist should be concerned with allowing the machine to cool.
90. **a.** 2,400 heat units are produced.
91. **b.** Bone has the highest atomic number.
92. **d.** Lead has the atomic number of 82.
93. **c.** The heating of the filament regulates mAs.
94. **d.** 1.35 should be added.
95. **d.** Rhenium is commonly used in combination with tungsten in the construction of an anode.
96. **c.** The wavelength of an x-ray beam is heterogenous.
97. **d.** Compton interaction affects contrast the most.
98. **b.** Characteristic radiation usually occurs after a Compton interaction.
99. **b.** With respect to magnification mode, the field of view is larger in the smallest setting.
100. **b.** Implementing the 15% rule will decrease heat units on an exposure.

IMAGE PRODUCTION AND EVALUATION

Key Terms

15% rule	cone
air-gap technique	crossed-hatch grid
anode heel effect	cylinder
aperture diaphragm	developer
automatic exposure controls	DICOM
characteristic curve	elongation
collimator	filter
compensating filter	fixer

Key Terms *continued*	
focal-spot blur	OID
focused grid	optical density
foreshortening	PACS
grid frequency	parallel grid
grid ratio	phosphor
grid	positive beam limitation
half value layer (HVL)	Potter-Bucky diaphragm
inherent filtration	radiographic contrast
intensifying screens	recorded detail
inverse square law	scatter radiation
kilovoltage peak (kVp)	shape distortion
law of reciprocity	SID/SOD
level	spatial distortion
low contrast	spatial resolution
magnification	speed
milliamperage per second (mAs)	window

Prime Technical Factors

Prime technical factors are factors under a technologist's control that influence the quantity (number of photons) and quality (penetrating ability of the photons) of the primary x-ray beam.

Milliamperage per Second

Referred to as *mAs*, it is attributed to the combination of milliamperage (mA) and time (seconds).

$$mAs = mA \times t$$

It represents the number of electrons emanated from the cathode that strike the target anode, creating x-ray photons over a period of time.

Kilovoltage Peak

Referred to as *kVp*, it is the maximum voltage applied to an x-ray tube, which in turn determines the energy of electrons traveling from cathode to target anode. This energy will determine the energy level of the x-ray photons created.

Distance

Distance generally refers to the distance from the x-ray source to the image receptor being imaged; however, it could be broken down into the distance from the source to the object being imaged, source to object distance (SOD), and the distance from the object to the image receptor, or source to image distance (SID).

$$SID = SOD + OID$$

$$SOD = SID - OID$$

$$OID = SID - SOD$$

Beam Quantity and Optical Density

Quantity (exposure or intensity) refers to the number of x-ray photons present in the primary useful beam. Technical factors influence quantity in varying ways.

Optical density (OD) refers to the overall degree of blackening on a radiograph, and in many ways this is directly proportional to quantity. Optical density can be measured by the amount of light transmitted through a processed film. The range of useful diagnostic densities measures 0.25 to 2.5 OD. Following is a discussion of how beam quality and optical density relate to mAs, kVp, and distance.

mAs

The quantity of x-rays within the primary beam is directly proportional to the mAs selected. If you double the mAs, the amount of x-ray photons emitted doubles. Milliamperage is the primary controller of optical density. The optical density of an image will remain constant as long as the mAs value is constant, regardless of the combination of mA and time utilized to get to that value. This is called the Law of Reciprocity. In order to visualize a perceptible change in density, mAs must be changed by 30%.

kVp

The quantity of the beam is affected by changes in kVp. When electron energy is increased in the tube, an incident electron can interact numerous times with different atoms in the target anode. Theoretically, the more electron interactions, the more x-ray photons of various energies could be produced, and vice versa. In the diagnostic range of kVp (40–150 kVp), the 15% rule has been adopted to represent the change to quantity. As the quantity fluctuates, it will affect the optical density of the image; however, it is not routinely used to manipulate density due to the changes in contrast that also take place, as outlined later.

- A 15% increase in kVp = double the quantity = double the density
- A 15% decrease in kVp = half the quantity = half the density

Consider the following in compensating for these changes in order to maintain density across an image: If the kVp is raised 15%, mAs must be decreased by half. If the kVp is decreased 15%, mAs must be doubled.

Distance

Distance affects quantity and optical density in the following ways:

- The farther away the source is from the object, the less photons would interact due to the beam's divergence from its source. This would lead to a decrease in density.
- The closer to the source the object is, the more photons would interact due to the close proximity not allowing the beam to diverge. This would lead to an increase in density.

Numerically, the x-ray intensity is inverse to the square of the distance. These changes are best represented by the inverse square law:

$$\frac{I_1}{I_2} = \frac{D^2{}_2}{D^2{}_1}$$

Scattered Radiation

It is important to mention scattered radiation because its production adds unwanted densities to an image. Scattered radiation is produced by the object (patient) as x-ray photons undergo Compton (outer shell) interactions and scatter from their original path. Scatter production is influenced by three factors to be discussed as the chapter proceeds: kVp, field size, and patient thickness.

Beam Quality and Radiographic Contrast

Quality refers to the penetrability of the primary beam. The higher the energy of an x-ray beam, the more penetrating in tissue the beam will be. Quality is measured utilizing half value layer (HVL). HVL refers to the thickness of absorbing material that is necessary in order to reduce the intensity of the beam to half of its original intensity. HVL increases with higher kVp.

Radiographic contrast is defined as the differences in optical densities on an image. Radiographic contrast is a combination of both image receptor contrast (screen/film combination and processing) and subject contrast (size, shape, and atomic number of subject and kVp used). Contrast, specifically subject contrast, has a correlation with beam quality. Images that present large differences between adjacent densities are termed *high* (or *short scale*) contrast images. Images that present little difference between adjacent densities are called *low* (or *long scale*) contrast images. The following technical factors affect beam quality and contrast, and it should be noted that mAs have no noticeable effect on contrast.

kVp

kVp most affects the penetrability of the beam, as it represents the amount of energy behind the emitted x-ray photons.

- High energy photons = more penetrating = high quality
- Low energy photons = less penetrating = low quality
- kVp is the primary technical factor used in the manipulation of radiographic contrast.
- Low kVp = high contrast images—great differences in adjacent densities, more black and white images
- High kVp = low contrast images—little difference in adjacent densities, grayer images

It should be noted that increases in kVp also increase the amount of Compton interactions taking place in the patient, thereby reducing contrast by introducing unwanted densities onto the image. When it is necessary to use these high kVp techniques, it is advisable to employ the usage of a grid or collimator (if applicable) or both in order to reduce the amount of scatter reaching the image receptor.

Distance

While the source-to-image distance (SID) in the diagnostic range has no effect on the penetrability or contrast, the object-to-image receptor distance (OID) does affect radiographic contrast. The air-gap technique, in which the image receptor is placed approximately 15 cm from the subject, increases the OID. With this technique, some scattered radiation that would normally decrease contrast would now scatter away from the receptor. This would increase the contrast of the image.

Image Distortion

The Subject

The size, shape, and thickness of the object being imaged all contribute to image quality. Thicker body parts attenuate the beam more than thinner body parts. Also, images present different attenuation properties within the same subject. This is due to the varying tissue densities and atomic numbers of parts present in the human body.

The lowest differential absorption occurs in air, water, fat, and soft tissue. The highest differential absorption occurs in bone.

Motion

Motion affects recorded detail, which is the actual degree of sharpness of the object recorded on the image. Motion, specifically voluntary motion by the object, leads to the greatest loss of recorded detail in images (motion blur); however, motion blur can effectively be limited by the use of low-time mAs techniques and thorough patient instruction.

Magnification (Size Distortion)

Magnification refers to the object in the image appearing larger than the actual object. When magnified, the image appears unsharp, especially around its borders. Therefore, it is general practice to keep magnification to a minimum in radiography. Distance most affects image magnification. It is necessary to use the largest possible SID while also employing the smallest possible OID to best reduce magnification. The magnification factor can be calculated in two ways. If the actual object size is known, then it can be calculated in comparison to the size of the object in the image. Or it can be calculated by comparing the source-to-image distance (SID) to the source-to-object distance (SOD):

$$MF = IS/OS$$

$$MF = SID/SOD$$

$$MF = IS/OS = SID/SOD$$

Shape Distortion

Shape distortion refers to an unequal magnification of different parts of the same object. This generally stems from the plane of the object being imaged and the angle of the central ray and/or the angle of the image receptor. Thicker objects generally appear more distorted than thinner objects due to the changing OID across thicker objects. If the object itself is not perfectly parallel with the plane of the image receptor, then the resulting image will display shape distortion. Foreshortening (object appearing smaller than actual) occurs when there is misalignment of the object. Elongation (object appearing longer than actual) occurs when the central ray is angled along the object.

In the body, multiple objects are generally positioned at differing OIDs throughout. Because of this, spatial distortion occurs that commonly results in either superimposition or nonvisualization of objects in a single projection.

Focal-Spot Blur (Penumbra)

Focal-spot blur occurs due to the fact that the x-ray beam does not originate from a single point source, but from the area known as the effective focal spot, which is rectangular in size. The blur appears on the periphery of the image and is generally more obvious on the side of the image relating to the cathode side of the tube. This adversely affects recorded detail and spatial resolution. It can be calculated as follows:

Focal spot blur = **(effective focal spot)** × **(OID/SOD)**

Anode Heel Effect

Due to the photon-absorbing capability of the anode, when photons are emitted there are more photons present on the cathode side of the x-ray tube. Therefore, there is more focal spot blur present on the cathode side. For certain examinations, this emission can be utilized to the technologist's advantage. An example of this would be for the lateral thoracic spine, where there are great subject differences between the shoulder portion of the spine and the lower spine. In this case, the shoulder portion of the subject should be placed on the cathode side of the tube to take advantage of the increased photon numbers.

The Grid

The grid is a device made up of thin strips of radiopaque material (usually lead) separated by a radiolucent interspace (made of aluminum or fiber) encased by a thin cover of aluminum. The cover is designed to effectively reduce the amount of scatter reaching the image receptor while allowing the x-rays whose path is straight to continue on their course (the remnant beam).

Grid Ratio

Grid ratio is calculated by the height of the lead strip divided by the interspace.

$$\text{Grid ratio} = h/D$$

High ratio grids offer a greater degree of scatter cleanup than low ratio grids; however, when utilizing a higher ratio grid, it is also necessary to increase the radiation exposure in order to compensate for the absorption.

Grid frequency: Refers to the number of grid strips per centimeter. As grid frequency increases, scatter cleanup increases due to the increase in grid strips and the smaller interspaces now associated; however, exposure must be increased due to the extra absorption.

Contrast improvement factor: Contrast improvement factor is a comparison made to determine the amount of contrast improvement a grid provides by comparing an exposure with a grid to an exposure without the grid of the same subject. It can be calculated as follows:

$$K = (\text{contrast with grid})/(\text{contrast without grid})$$

Contrast improvement is best seen with higher ratio grids.

Typical grid improvement factors:

No grid = 1

5:1 = 2

6:1 = 3

8:1 = 4

12:1 = 5

16:1 = 6

While it is true that contrast is improved, it should be noted that density is also affected when a grid is introduced. Due to its absorption properties, some of the primary beam will always be absorbed by the strips. In order to compensate for this loss of density, mAs must be adjusted accordingly. Utilizing the improvement factors, this can be calculated using the following:

New mAs = old mAs × (new grid factor/old grid factor)

Grid Types

Parallel: All grid strips are parallel to each other and easy to manufacture. However, grid cutoff, which refers to the primary beam being stopped by the grid and kept from reaching the image, is an issue due to the position of the strips in relation to the divergence of the beam.

Crossed: In crossed grids, two parallel grids are interlaced together so that the strips of each grid are positioned perpendicular to one another. They are more efficient then parallel grids; however, there is significant cutoff if the central ray is not perfectly centered or the tube is angled in any way.

Focused: Focused grids are constructed so that the grid strips are angled to coincide with the divergence of the x-ray beam from its source at a given distance, which should be noted on the grid itself.

Misalignment: When a focused grid is misaligned, various quality issues take place.

- Off level: improperly positioned tube or tilted grid which presents cutoff throughout the image.
- Off center: laterally improperly positioned tube which presents cutoff throughout the image.
- Off focus: tube positioned at the nonfocus distance, cutoff toward the periphery, depending on the distance.
- Upside-down: severe cutoff on the periphery.

Grid Motion

Since the strips are radiopaque, they do present themselves on an image, especially when low frequency grids are employed. The Potter-Bucky diaphragm assists in the motion of a grid to blur out the visibility of these strips. The types of motion used are as follows:

- Reciprocating: back and forth motion utilizing a motor during exposure.

- Oscillating: utilizes an electromagnet to pull the spring-attached grid to one side, releasing it upon exposure and allowing movement in a circular fashion.

Filtration

The purpose of x-ray filtration is to remove low-energy photons which serve little diagnostic purpose and contribute to an increased patient dose. The end result is a higher quality x-ray beam. Aluminum serves as the most common filtering material.

Inherent Filtration

Any filtration pertaining to the glass envelope, metal enclosure, or dielectric oil surrounding the tube is considered inherent filtration.

Added Filtration

This refers to the thin piece of aluminum between the housing and the collimator, as well as the mirror associated with collimator boxes.

Compensating Filter

Compensating filters are used to compensate when there are varying thicknesses in the part being imaged in order to produce a more uniform density across the image.

Wedge Filter

Wedge filters are commonly used with parts like the foot where the thicker portion of the wedge would be positioned to correspond with the thinnest portion of the part being examined.

Trough Filter

Trough filters are commonly used in chest imaging with the thin portion of the wedge over the center of the chest and the thicker portion over the lungs.

Total Filtration

Total filtration equals added filtration plus inherent filtration.

Beam Restriction

As field size increases, scatter radiation reaching the image receptor increases, thus leading to reduced contrast across the image. Therefore, it is best to reduce field size by restricting the beam. When field size is decreased to improve contrast, a lower density presents itself that must be compensated for by increasing exposure. A good radiographer only images the object being examined, utilizing whatever beam restrictors that are available. There are different types of beam restrictors:

Aperture diaphragm: A simple, flat, lead diaphragm with an opening cut into its center that mounts onto the tube head. It is only useful for specific exams.

Cone and cylinder restrictors: These are extensions built onto an aperture diaphragm that either extends straight (cylinder) or flares (cone) to its target. They are only useful for specific exams.

Collimator: This is a device that allows for unlimited restriction sizes, has filtering properties, and provides a light source to assist the technologist in positioning. It consists of two sets of shutters positioned on different levels. One pair is located in relation to the tube head; this set assists in minimizing off-focus radiation from impairing the image. The second pair is positioned lower. A small light source is located off to the side of the box, and the light is reflected off a mirror positioned to correspond to the path of the primary beam.

Positive beam limitation (PBL): PBL automatically adjusts the collimation to correspond with the size of the film loaded. It will restrict it to that size or smaller.

Technique Charts

Technique charts were developed in order to assist the technologist in utilizing standardized techniques to produce consistent, optimum images. A technique chart is created using only calibrated equipment and the entire imaging system (screen/film, CR, grid type, standardized SID, etc.). There are various types of technique charts:

Variable kVp: Fixed mAs are used for the body part. kVp is modified by utilizing a measuring tool known as a caliper and by increasing/decreasing by 2 kV per cm thickness. Short-scale contrast can be obtained, however dosages are normally higher, and there is a general decrease in exposure latitude.

Fixed kVp: kVp is fixed for the specific body part imaged, however mAs are modified. Parts are grouped according to size, and the mAs are adjusted based on a 30% increase/decrease from the average size. High kVps are utilized whenever certain conditions are present.

Contrast media: Barium requires techniques above 100 kVp.

Chest radiography: This commonly requires techniques above 120 kVp in order to adequately represent the various tissue densities present in the chest.

Pathology: Certain pathologies require a variation to the common techniques used. They are grouped as destructive (tissues are broken down leading to a reduction in that tissue's density) or additive.

Destructive pathologies appear radiolucent and require a decrease in techniques. Examples include:

- pnuemothorax
- osteoporosis
- myeloma
- emphysema
- bowel obstruction

Additive pathologies appear radiopaque and require increases in techniques. Examples include:

- ascites
- atelectasis
- cirrhosis
- tumors
- pneumonia
- pleural effusion

Automatic Exposure Controls

Automatic exposure controls contain three specifically positioned ionization chambers that serve to terminate an exposure when the system has determined that a certain measured exposure reaches the image receptor. Each chamber selected gets saturated by the remnant beam and, as more radiation is soaked up, ions will charge the chamber(s) until a preset value is attained. The exposure is then terminated. The technologist can manipulate the kVp utilized as well as the mA, if this is desired. Time, however, will be determined by the AEC, and cannot be adjusted. It is calibrated to maintain a certain optical density for each examination.

The technologist also has the control over which ionization chambers will be available for the exposure. Positioning must be accurate in relation to the chambers selected. If it is not, it could result in the quick termination of exposure if the part is not positioned over the selected cell, resulting in an image with a lack of density. Safeguards are also in place to ensure that over-dosage does not take place. The system contains a backup timer, which will terminate the exposure in the event that the AEC does not terminate at the programmed density for that exam. The device could also be manipulated to allow for more or less absorption by the chambers per cell. The +1, +2, −1, −2 settings, which can be selected by the technologist prior to exposure, allow for less (−1, etc.) or more (+1, etc.) saturation to compensate for part thickness, if necessary.

Intensifying Screens

Intensifying screens are used in conjunction with specific types of film in order to utilize light photons created by the screen to assist in lowering the dosage necessary in the creation of a diagnostic image. The film itself must be in complete contact with the intensifying screen in the cassette in order to get the most optimal result. Some cassettes have intensifying screens on both sides in order to accommodate double emulsion film. The main drawback is that despite its main advantage, there is a loss of detail when compared to a direct exposure onto film. Spatial resolution refers to a screen's ability to produce a clean, clear image.

Construction

Screens consist of an active layer known as a phosphor attached to a base. In between these parts is a reflective layer, which helps direct more light photons toward the film.

The base is polyester, which provides a stable surface. Phosphor is a material that will emit light photons when stimulated by radiation. It is this light that will help quicken the exposure of a film. The more light a certain type of phosphor emits when exposed by a given x-ray exposure, the faster the speed of the screen. The rare earth phosphors (gadolinium, yttrium, and lanthanum) are considered the most common of the fastest screens. Light-absorbing dyes are sometimes added to help in the control of the dispersal of light by the phosphor. Phosphors should also only emit light when stimulated. Any light that is emitted after exposure has ceased (afterglow) is detrimental to the film, as it will continue to undergo exposure.

Speed

A screen's speed is based on its ability to convert incoming remnant x-ray photons into a number of light photons, which is also known as conversion efficiency. It is described numerically as par = 100, and fast = 400. The more light photons created by one photon, the higher the conversion efficiency. Thicker phosphors have higher speeds than thinner ones; however, there is a limit to speed because if a phosphor is too thick, photons will be unable to penetrate deep enough for light to be emitted. Speed directly influences detail. Slower speed screens display more detail then high-speed screens.

Screen speed directly affects density: The higher the speed, the greater the density on the film. In order to compensate for a possible change in screen speed, use the following formula:

New mAs = old mAs × (old screen speed/new screen speed)

Spatial Resolution

Spatial resolution refers to the level of detail. It also refers to how small an object can be and still appear on an image. While the focal spot size is the primary control of the actual object size that can be imaged, the image is being created by light photons emitted by the screen, which creates blur, which degrades the image. The higher the speed, the more blur, and the less the spatial resolution. Slower speeds have higher spatial resolution.

The Image Receptor

Film and film screens represent the most basic and common receptors in radiography.

Film

Film consists of two main parts: the base and the emulsion. Between the base and emulsion is an adhesive layer which helps stick the emulsion to the base. There is a supercoat (or overcoat) layered on top of the emulsion to provide some protection from some mishandling.

Base: Made of polyester, the base provides a rigid, stable surface for the emulsion to be coated on and is tinted blue in order to ease eye strain of the radiologist.

Dimensional stability: This is the ability of the base to maintain its size and shape during processing.

Emulsion: The interactive part of film, it consists of a mixture of gelatin and silver halide crystals.

Gelatin: This is clear and porous in order to allow for the transmission of both light and the processing chemicals.

Silver halide crystals: These are a combination of 98% silver bromide and 2% silver iodide. Imperfections, known as sensitivity specks, in the crystals are the areas where the latent or invisible image will form after interaction with light and/or x-ray.

Film latent image formation: The latent image, as best explained by the Gurney-Mott theory, is produced when an incident photon strikes the silver-halide crystals. Silver ions are produced as photons knock electrons out of their original orbits. These silver ions migrate to the sensitivity specks, neutralize there, and form deposits, resulting in the latent image. After processing chemicals are introduced to an exposed film, it becomes a manifest, or visible, image.

Contrast: Refers to the contrast inherent in the film; it will vary, which directly affects the contrast of the finished image. Smaller grain film is higher contrast film, while larger grain film is lower contrast.

Speed: This also varies depending on construction. Film speed is defined as the sensitivity of film to light and x-ray. Grain size and shape also define the speed of a specific film. Small grain film is considered lower speed, while larger grains are considered higher speed. This directly influences exposures, as higher speed film require less exposure to provide a diagnostic image than slower speed film. Double emulsion film has emulsion coating on both sides of the base, thus providing double the speed a single emulsion would provide.

Spectral matching: The most important consideration in spectral matching is the ability of a film to react to the color of light emitted by the intensifying screen it is paired with. If improperly matched, speed will be impaired severely, which will lead to unnecessary increases in exposures.

Quantum mottle: Quantum mottle occurs when too few x-ray photons strike the image receptor and it appears as graininess throughout the image.

Processing: Processing is the process where the invisible latent image is converted into the visible manifest image.

Screen Film

Screen film is designed to be used in conjunction with an intensifying screen, depending on the size and layout of the silver halide crystals during construction.

Direct Exposure Film

Direct exposure film is a high-detail film that is directly exposed only by x-ray; there are no intensifying screens utilized. This type of film requires an extremely high dose in order to get a diagnostic exposure and should only be used with thin body parts.

Laser Film

Laser film is red-light sensitive film and must be handled in total darkness.

Mammography Film

Mammography film is a single emulsion film to be utilized with a single intensifying screen. An antihalation coating on the back helps to prevent reflected light from the cassette backing from affecting the film.

Digital Imaging

Digital imaging is a more recently introduced imaging system that utilizes computer technology and allows for wider latitudes in regard to technique. This helps reduce unnecessary exposures, and it allows post-processing manipulation of images. It generally involves an exposure onto either an arranged set of flat panel radiation detectors (direct radiography, or DR) or onto a photostimulable phosphor (computed radiography, or CR), which is then converted into a matrix by a computer. The phosphors in CR trap the energy created after the exit radiation strikes the phosphor. When the plate is put through the processing unit or scanner, a laser beam causes the emission of this energy, which is then detected by a photo-multiplier tube. This information is then converted from analog into a digital format by an ADC converter. Image plates can store information for several hours until the energy stored slowly starts to fade.

Image Matrix

An image matrix is made up of pixels, or cells, arranged in rows and columns that are given a certain numerical value based on the information given. Each pixel corresponds to a particular shade of gray, which is based on the original energy level from a particular area of the subject. This is called a voxel. The spatial resolution of a digital image is determined by its image matrix. A histogram is generated to allow for the visualization of the range of pixel values. Each specific exam will have a specific algorithm built into the software that will allow for reconstruction of the correct histogram values.

Digital Spatial Resolution

Digital system spatial resolution (visualization of the separation of two objects) is much finer when utilizing DR equipment than with CR equipment. However, the drawback is that the resolution of the image will always be based upon the resolution of the monitors presenting that image.

Film Storage and Handling

Film must be stored in a cool, dry, and dark place away from radiated materials. It should be stored on edge to prevent pressure artifact. The temperature of a storage area should be lower than 68 degrees F, and the best storage mechanism is refrigeration. At higher temperatures, the film becomes susceptible to fog, a severe loss of contrast along the film. The humidity should be between 40% and 60%. If the humidity is too high, then there is a risk of fog. If it is too dry, then the possibility of static artifacts exists. Film also has an expiration date provided by the manufacturer. Film must always be handled in the dark, as light would fog a film. When using screen film, a safelight is generally present to provide a source of illumination for the handler. Depending on the type of film utilized, the filter of the safelight would change. For blue film, an amber filter is used (Wratten 6B being the most common). For green film, a red filter (Kodak GBX) is used. This safelight should be 5 feet or higher from the work surface, and the lightbulb should be no greater than 15 watts. Film should be handled gently and with clean hands. Artifacts can be generated due to mishandling.

Chemical Solutions and Processing

Each film must go through a series of solutions in order for the image to manifest.

Developer: The developer is an alkali solution that reduces the silver ions present at the sensitivity speck into black metallic silver. As the gelatin swells during this process, the shades of gray are produced quickly, while the overall blackness of the film proceeds more slowly. It must be kept at 95 degrees F. If too hot, fog will take place. If too cold, the film will appear too light.

Fixer: An acidic solution which stops the developing process, removes the unexposed silver bromide crystals from the film, and preserves and hardens the film.

Wash: Water that washes away all residual chemicals.

Dry: Warm air blown onto the film in order to facilitate drying.

Automatic processor: The common automatic processor takes 90 seconds, and its main advantage is consistent processing from film to film. The solutions are contained within wells. There are various systems working in conjunction with one another in order to maintain the consistency. The processor should be maintained daily. Upon startup, the feed tray and crossover rollers should be wiped clean of any dust and debris that may have collected, sensitometry should be performed, and the system should be given enough time to get to temperature. Upon shutdown, the panel over the chemicals should be left open 1 inch in order to allow moisture to escape.

Transport system: The transport system starts at the feed tray. It includes a system of rollers that are motor driven, which moves the film through the processor. The rate of movement controls the amount of time each film remains in the solutions. The rollers in the solution are aligned in racks within the tanks, which enable them to fully immerse the film into the solution. Guide shoes are positioned to ensure the direction of travel. At the bottom of the rack assembly is a turnaround assembly that turns the film back toward the exit of the solution. Out of the solution are crossover racks that guide the film into the next well.

Temperature control system: The temperature control system maintains the temperature of the solutions.

Circulation system: The circulation system agitates the solution in order for the chemicals to maintain their activity.

Replenishment system: The replenishment system is activated at every 14 inches of film that enters the processor. It ensures fresh solution is always readily available in the wells.

Dryer system: The dryer system utilizes air blowers that dry the film.

Artifacts

Processor-Related Artifacts

Guide shoe marks: Found on the leading and trailing edges and follow the direction that the film travels.

Pi lines: These are from dirty rollers and can occur every 3.14 inches.

Pick-off marks: Occur due to excessive dirt on rollers; emulsion is picked off.

Chemical fog: Gives a discolored curtain effect.

Handling-Related Artifacts

Light or radiation fog: Appears as an increased density across the exposed areas.

Kink marks: Appear as a zigzag along the horizontal axis that shows the histogram does not start at the beginning; it may result from undue stress to the paper or film before processing.

Sensitometry

Sensitometry is the study of the relationship between the exposure and the density after processing. It is best perceived utilizing the characteristic curve, or Hurter and Driffeld Curve. The curve can be developed utilizing a step wedge and a tool known as a densitometer, which measures the optical density on a film. The step wedge is placed on a cassette, an exposure is taken, the film is processed, and the step wedge pattern is measured and recorded on a graph.

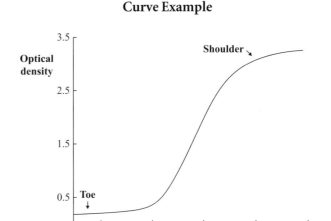

Curve Example

As mentioned before, the useful range of OD is 0.25–2.5. On a curve, the toe portion represents underexposure, the straight line portion represents the useful densities on the film, while the shoulder represents overexposure. The difference from step to step on the straight line portion is measured as log-relative exposure. The slope of the curve shows a variety of information regarding the film system beyond just density. Speed can be determined by the slope of the curve; the steeper the slope, the faster the film. Contrast can be ascertained by the slope of the curve; the steeper the slope, the higher the contrast. Technique latitude follows the same principal: the steeper the slope, the lower the latitude.

Electronic Picture Archiving and Communication Systems (PACS)

PACS was developed to provide digital image storage. It provides the rapid retrieval of images and allows for simultaneous access at multiple sites. The format most commonly utilized is called DICOM, or Digital Imaging and Communications in Medicine. Images can be stored on hard disks both locally and/or remotely in another facility, and they are easily accessed. PACS interfaces with both a hospital information system and radiology information system. The HIS provides patient information and order data to the RIS, which allows technologists and radiologists a much more simplified way to access and enter data. Each image's data set in PACS contains the histogram associated with each individual image, which is a graphical representation of the energy distribution of the pixels represented in an image. The x-axis represents the blackness of the image, and the y-axis represents the size of the area containing that energy. This allows for manipulation along the histogram of the contrast presented on the image. The x-axis of the image histogram can be adjusted by window width so that more of the gray tones stand out in the image. By window-leveling up and down the y-axis, this will allow for more or less brightness in the image.

CHAPTER 7 REVIEW QUESTIONS

1. Which of the groups of exposure factors will produce the greatest radiographic density?
   a. 100 mA, 0.30 sec
   b. 200 mA, 0.10 sec
   c. 400 mA, 0.03 sec
   d. 600 mA, 0.03 sec

2. A grid is usually employed in which of the following circumstances?
   1. when using 80 kVp
   2. when the patient is obese
   3. when a lower patient dose is required

   a. 1 only
   b. 3 only
   c. 1 and 2 only
   d. 1, 2, and 3

3. Exposure factors of 90 kVp and 4 mAs are used for a particular nongrid exposure. What should be the new mAs if an 8:1 grid is added?
   a. 8 mAs
   b. 12 mAs
   c. 16 mAs
   d. 20 mAs

4. In a PA projection of the chest being used for cardiac evaluation, the heart measures 15.2 cm between its widest points. If the magnification factor is known to be 1.3, what is the actual diameter of the heart?
   a. 9.7 cm
   b. 11.7 cm
   c. 19.7 cm
   d. 20.3 cm

5. Which of the following pathologic conditions would require a decrease in exposure factors?
   a. empyema
   b. osteoporosis
   c. pneumonia
   d. ascites

6. Which of the following phrases are true when comparing a high ratio grid to a low ratio grid?
   1. absorbs more scatter radiation
   2. absorbs more of the primary beam
   3. requires a high exposure technique

   a. 1 only
   b. 1 and 2 only
   c. 1 and 3 only
   d. 1, 2, and 3

7. The function of the automatic processor's transport system is to
   a. allow for proper replacement of solution.
   b. monitor the temperature.
   c. provide the proper agitation of solutions.
   d. move the film and change its direction.

8. Which of the following is most likely to produce a radiograph with a long scale of contrast?
   a. increased kVp
   b. decreased screen speed
   c. increased mAs
   d. increased SID

9. Which of the following is NOT related to recorded detail?
   a. OID
   b. film speed
   c. using a 0.1 focal spot
   d. mAs

10. An mAs value will reproduce the same optical density on a radiograph regardless of the combination of mA and time utilized. This describes the
   a. Law of Bergonie and Tribondeau.
   b. line focus principle.
   c. law of reciprocity.
   d. grid ratio.

11. The term *spectral matching* refers to the fact that film sensitivity must be matched with the film's
   a. speed.
   b. contrast enhancement factor.
   c. phosphor color fluorescence.
   d. histogram.

12. In order to minimize the magnification of an object, if OID cannot be minimized sufficiently, what can a technologist do?
   a. utilize greater source-to-image distance
   b. use high speed film
   c. utilize a breathing technique
   d. take the image as is

13. Which of the following groups of exposure factors would be most appropriate to control involuntary motion?
   a. 600 mA, 0.12 sec
   b. 300 mA, 0.2 sec
   c. 1,000 mA, 0.02 sec
   d. 400 mA, 0.25 sec

14. Due to the effects of the absorbing ability of the anode, when x-ray photons are created more photons are emitted by the tube
   a. on the cathode side of the beam.
   b. on the anode side of the beam.
   c. in the center of the primary beam.
   d. in equal distribution.

15. Which of the following technical factor(s) could possibly be used to manipulate optical density of an image?
   1. milliamperage
   2. exposure time
   3. kilovoltage
   a. 1 only
   b. 2 only
   c. 1 and 3 only
   d. 1, 2, and 3

16. In x-ray production, the term *quantity* typically refers to
   a. kVp.
   b. SOD.
   c. mAs.
   d. OID.

17. When a body part exhibits unequal thicknesses throughout, what could best be utilized in order to present an even image?
   a. a grid
   b. intensifying screens
   c. a compensating filter
   d. a caliper

18. Which of the following groups of exposure factors will produce the shortest scale of contrast?
   a. 800 mA, 0.02 sec, 70 kVp
   b. 100 mA, 0.3 sec, 85 kVp
   c. 600 mA, 0.025 sec, 90 kVp
   d. 800 mA, 0.01 sec, 80 kVp

19. Which describes the height of a grid's lead strips compared to the distance between each strip?
   a. grid ratio
   b. focusing distance
   c. frequency
   d. conversion efficiency

20. A satisfactory radiograph of the abdomen was made at a 42-inch SID using a 300 mA, 0.06 second exposure and 80 kVp. If the new distance is changed to 38 inches, what is the new mAs required?
    a. 10 mAs
    b. 15 mAs
    c. 20 mAs
    d. 24 mAs

21. Radiographic contrast can be defined as
    a. graininess throughout the image.
    b. parallel lines running along the periphery.
    c. difference in the optical densities between adjacent areas on a radiograph.
    d. overall blackening on the image.

22. Any unwanted marks that appear on a film after it is processed are called film
    a. glitches.
    b. lines.
    c. artifacts.
    d. scatter.

23. Which grid ratio would present the greatest cleanup of scattered radiation?
    a. 6:1
    b. 8:1
    c. 12:1
    d. 16:1

24. The primary exposure factor that is used for regulating radiographic contrast is
    a. SID.
    b. focal spot size.
    c. kVp.
    d. mAs.

25. The total amount of diagnostic filtration can be calculated by finding the sum of
    a. inherent and compensating filtration.
    b. inherent filtration.
    c. added and compensating filtration.
    d. inherent and added filtration.

26. Which of the following is (are) the major advantage(s) of collimators over other common beam-limiting devices?

    1. the variety of field sizes available
    2. a light
    3. better cleanup of scatter radiation

    a. 1 only
    b. 1 and 2 only
    c. 1 and 3 only
    d. 1, 2, and 3

27. If a film is mishandled prior to processing, it is likely that the film will exhibit areas of
    a. tree static artifact.
    b. starry night artifact.
    c. minus density artifact.
    d. plus density artifact.

28. Of the following groups of exposure factors, which will exhibit the greatest density on a film?
    a. 300 mA, 0.08 sec, 8:1 grid ratio, 72 inches
    b. 200 mA, .25 sec, 8:1 grid ratio, 72 inches
    c. 300 mA, 0.08 sec, 8:1 grid ratio, 40 inches
    d. 300 mA, 0.08 sec, 8:1 grid ratio, 36 inches

29. In an automatic processor, the function of the developer is to
    a. remove all unexposed silver halide crystals.
    b. change the manifest image into a latent image.
    c. wash the film.
    d. serve as the sensitivity specks.

30. Which of the following does NOT contribute to radiographic contrast on an image?
   a. atomic number of target
   b. kilovoltage used
   c. mAs used
   d. use of a radiographic grid

31. Which of the following would NOT be a reason to utilize a high ratio grid?
   a. absorption of scattered radiation
   b. when a low dose technique is required
   c. to present a short scale of contrast
   d. for an extremely thick body part

32. In the characteristic curve of a specific type of film, what does the toe portion represent?
   a. overexposure
   b. underexposure
   c. emission spectrum
   d. no exposure

33. Which of the following factors is (are) necessary in order for quantum mottle to be present on an image?

   1. fast speed screen
   2. high kVp technique
   3. low mA technique

   a. 1 only
   b. 1 and 2 only
   c. 1 and 3 only
   d. 1, 2, and 3

34. Which of the following exposure techniques will produce the shortest scale of contrast?
   a. 200 mA, 0.75 sec, 52 kVp
   b. 400 mA, 0.60 sec, 58 kVp
   c. 300 mA, 0.10 sec, 65 kVp
   d. 800 mA, 0.05 sec, 90 kVp

35. Magnification and foreshortening are considered
   a. size distortion.
   b. shape distortion.
   c. film distortion.
   d. average gradient.

36. The source-to-image distance and the size of the focal spot have the greatest effect on which radiographic property?
   a. contrast
   b. density
   c. recorded detail
   d. shape distortion

37. The overall blackening present on a radiograph is the definition of
   a. density.
   b. contrast.
   c. magnification.
   d. emission spectrum.

38. An advantage of digital imaging is
   a. no need for a grid.
   b. lessened filtration is necessary.
   c. slower processing times.
   d. greater exposure latitude.

39. Which of the following groups of exposure factors would produce the greatest radiographic density?
   a. 10 mAs, 60 kVp, 8:1 grid ratio
   b. 20 mAs, 60 kVp, 16:1 grid ratio
   c. 10 mAs, 74 kVp, no grid
   d. 10 mAs, 74 kVp, 6:1 grid

40. If a 40-inch SID is utilized on an object located 4 inches from the image receptor, what will the magnification factor be?
a. 1
b. 1.1
c. 1.5
d. 2

41. Changing the AEC setting from 0 to +1 would have what effect on density?
a. none
b. 10% increase
c. 25% decrease
d. 25% increase

42. What is the primary advantage of a moving grid over a stationary grid?
a. blurring of the grid strips
b. better scatter cleanup
c. ease of portable use
d. better cleanup of primary beam

43. Which of the following factors would produce the film with the longest scale of contrast?
a. 200 mA, 0.1 sec., 70 kVp, 6:1 grid ratio
b. 100 mA, 0.2 sec, 70 kVp, 8:1 grid ratio
c. 200 mA, 0.1 sec, 70 kVp, 12:1 grid ratio
d. 100 mA, 0.2 sec, 70 kVp, 16:1 grid ratio

44. In order to create a variable kV technique chart, kV should be adjusted based on what change in patient thickness?
a. 1 cm
b. 2 cm
c. 3 cm
d. 4 cm

45. Films that are pulled from the film bin quickly in humid conditions run the risk of producing
a. fog.
b. minus density.
c. static artifact.
d. pick-off.

46. Any material that emits light upon stimulation from a type of radiating photon is called a(n)
a. electron.
b. phosphor.
c. image receptor.
d. filter.

47. The type of distortion that occurs when the primary beam is angled against the long axis of the part is called
a. magnification.
b. foreshortening.
c. elongation.
d. flux gain.

48. kVp relates to a beam's
a. energy level.
b. quality.
c. penetrating power.
d. all of the above.

49. When compensation filtration is present, what is the effect on the finished image?
a. decrease in skin dose
b. increase in contrast along the part
c. decrease in contrast along the part
d. none of the above

50. When a radiographic grid is utilized, the image will appear with
a. increased contrast.
b. increased density.
c. decreased contrast.
d. none of the above.

51. In order to best utilize the anode heel effect, when radiographing the thoracic spine in a lateral position, the shoulder portion should be placed
   a. in the center of the primary beam.
   b. on the anode end of the tube.
   c. on the cathode end of the tube.
   d. on either end of the tube.

52. When performing a study, if the SID must be increased by double the original distance, what effect will this have on density?
   a. Density would double.
   b. Density would be halved.
   c. Density would be reduced by a factor of 4.
   d. Density would be the same.

53. The invisible image present on an exposed film is called the
   a. manifest image.
   b. latent image.
   c. double exposed image.
   d. post-processed image.

54. The type of computed radiography imaging plate most commonly used is
   a. barium platinocyanide.
   b. rare earth phosphors.
   c. photostimulable phosphors.
   d. cesium iodide plates.

55. The range of acceptable optical density is
   a. 0.01–0.025 OD.
   b. 0.25–2.0 OD.
   c. 2.0–2.5 OD.
   d. 2.5–5.0 OD.

56. Which information can be discerned from a characteristic curve?
   1. Speed
   2. Latitude
   3. Contrast

   a. 1 only
   b. 1 and 2 only
   c. 2 and 3 only
   d. 1, 2, and 3

57. The replenishment system in the automatic processor will initiate
   a. when 17 inches of film pass.
   b. when 14 inches of film pass.
   c. when 10 inches of film pass.
   d. constantly.

58. What is the magnification factor if an object 3 inches in height is imaged from 36 inches?
   a. 0.9
   b. 1
   c. 1.1
   d. 1.5

59. If an exam is performed at 10 mAs utilizing a 200 speed screen, what would the new mAs be if the exam was performed utilizing a 400 speed screen?
   a. 5 mAs
   b. 10 mAs
   c. 20 mAs
   d. 40 mAs

60. If a lower speed film is utilized, what would be the effect on density (assuming no factors are changed)?
   a. Density would increase.
   b. Density would decrease.
   c. Density would remain the same.
   d. Density would increase by a factor of 4.

61. A grid error in which severe cutoff along the periphery of the image is present is due to
a. the grid being off focus.
b. the grid being off level.
c. the grid being upside down.
d. the grid being off center.

62. What would be a characteristic of a low contrast film?
a. large grain
b. small grain
c. green tinted
d. self-processing

63. The shoulder portion of the characteristic curve represents
a. overexposure.
b. underexposure.
c. average gradient.
d. diagnostic range.

64. In comparison to high speed screens, low speed screens require which of the following?
1. lower exposure
2. higher exposure
3. greater detail

a. 1 only
b. 1 and 3
c. 2 and 3
d. 1, 2, and 3

65. Utilizing a type of beam restrictor will have what effect on radiographic contrast?
a. It increases.
b. It decreases by a factor of 2.
c. There is no effect.
d. It decreases proportionally.

66. Which system in an automatic processor provides fresh chemistry?
a. transport
b. recirculation
c. wash/dry
d. replenishment

67. When utilizing an AEC, which factors can be manipulated by a technologist?
1. kVp
2. mA
3. time

a. 1 only
b. 2 only
c. 1 and 2 only
d. 1, 2, and 3

68. If a part is not properly positioned over an ionization chamber, the resulting image will appear
a. overpenetrated.
b. underpenetrated.
c. with no discernable effect.
d. grainy.

69. A graininess that appears on a high speed film when a low mA station is selected is called
a. a penumbra.
b. a quantum mottle.
c. a Wye pattern.
d. static.

70. Which of the following is NOT an example of a destructive pathology that requires a decrease in the technique used?
a. osteoporosis
b. empyema
c. emphysema
d. pneumothorax

71. What should be the minimum kVp utilized when imaging a patient with barium contrast?
a. 70
b. 80
c. 90
d. 100

72. In order to enhance both contrast and density when imaging a foot, which type of compensating filter should be used?
a. wedge
b. trough
c. inherent
d. step

73. In order to calculate an H and D Curve, a device called a _____ must be used to measure the density of each step.
a. sensitometer
b. penetrometer
c. densitometer
d. step wedge

74. The emulsion present in radiographic film is made up of both gelatin and
a. barium platinocyanide.
b. silver halide.
c. cesium iodide.
d. gadolinium.

75. Film should be stored below which temperature?
a. 98 degrees F
b. 100 degrees F
c. 68 degrees F
d. 80 degrees F

76. Any unwanted mark on a processed radiograph is referred to as
a. fog.
b. an artifact.
c. moiré.
d. noise.

77. Certain techniques/devices that could be used in order to limit the scatter reaching the IR include which of the following?

1. grid
2. anode heel
3. air gap

a. 1 only
b. 1 and 2
c. 1 and 3
d. 1, 2, and 3

78. Filtration serves to
a. focus the primary beam.
b. reduce low energy electrons.
c. cone the primary beam to a specific area.
d. increase skin dosage.

79. The degree of sharpness of the object recorded on the image defines
a. spatial resolution.
b. recorded detail.
c. penumbra.
d. filtering effect.

80. The image being adjusted so that it allows more of the gray tones to stand out describes
a. window width.
b. window level.
c. mAs.
d. spatial resolution.

81. Which of the following will NOT affect patient dose?
    a. vmilliamperage per second
    b. filtration
    c. focal spot size
    d. kilovoltage peak

82. During automatic processing, the replenishment rate of the chemicals associated is determined by which factor?
    a. size of film
    b. temperature of processor
    c. speed of the roller system
    d. type of chemicals used

83. Which of the following would be the result of an exam done utilizing insufficient kVp?
    a. over-penetration
    b. excessive detail
    c. under-penetration
    d. quantum Mottle

84. In relation to recorded detail, what effect would an increase in the object-to-image distance cause to the image?
    a. Detail would not be affected.
    b. Detail would decrease inversely.
    c. Detail would increase proportionally.
    d. Detail would increase inversely.

85. In automatic processors, what is the primary purpose of the rollers?
    a. They transport film throughout processor.
    b. They recirculate chemistry throughout processor.
    c. They heat up the film sufficiently to assist in drying.
    d. They collect excess, unexposed black metallic silver.

86. Which effect would an image have if there was an increase in the photon energy in an x-ray beam?
    a. increased contrast
    b. decreased contrast
    c. decreased density
    d. no effect

87. An image of the knee was obtained utilizing 10 mAs, 60 kVP, and an 8:1 grid with 40 inches of source-to-image distance. If the image was then to be obtained utilizing 20 mAs with all other factors remaining the same, the end result would show
    a. increased density.
    b. decreased density.
    c. decreased contrast.
    d. increased magnification.

88. When utilizing a screen-film system, if the film is green sensitive and is paired with the improper screen, the end result will be
    a. increased contrast
    b. decreased contrast
    c. no effect on contrast
    d. diagnostic film

89. In order to produce a change equaling double the original density on an image, the kVp must be manipulated at least
    a. 5%.
    b. 15%.
    c. 20%.
    d. 30%.

90. Which of the following will allow for elongation of a part of an image?

    1. angling the x-ray tube to the part
    2. angling of the tube to the IR
    3. angling of the grid to the IR

a. 1 only
b. 2 only
c. 3 only
d. 1, 2, and 3

91. Which of the following factors may contribute to an increase in recorded detail on the finished image?

    1. large focal spot
    2. smaller focal spot
    3. slow screen film

a. 1 only
b. 2 only
c. 2 and 3 only
d. 1 and 3 only

92. If 20 mAs was utilized for a particular exposure with a par (200) speed screen, what would the new mAs be if a rare earth (800) speed screen was utilized?
a. 5
b. 10
c. 15
d. 20

93. The terminology *short-scale contrast* generally refers to a film that contains
a. few densities with great differences between them.
b. a large amount of densities with less differences between them.
c. the same density present throughout the entire image.
d. no density present throughout the entire image.

94. In order to properly test screen-film contact, which type of exam is performed?
a. spin top
b. dosimetry
c. wire mesh
d. kV calibration

95. In an automatic processor, what is the purpose of the crossover rollers?
a. to direct a film from the bottom of the solution tank back toward the top
b. to direct a film into the processor
c. to direct a film through the dryer
d. to direct a film from one solution into the next

96. If the kilovoltage had to be increased in order to assist in the reduction of patient dosage, the most noticeable effect on the finished image would be
a. decreased contrast.
b. increased contrast.
c. increased light fog.
d. decreased density.

97. What would be the mA if 0.25 seconds were used in order to produce 100 mAs?
a. 200 mA
b. 400 mA
c. 600 mA
d. 1,000 mA

98. What is the mAs value if 800 mA was used at 0.05 seconds?
a. 10
b. 20
c. 30
d. 40

99. A grid where the grid strips are angled to coincide with the divergence of the x-ray beam from its source at a given distance is called
a. cross-hatched.
b. parallel.
c. focused.
d. oscillating.

100. The visualization of the separation of two objects best describes
a. penumbra.
b. spatial resolution.
c. focal spot blur.
d. elongation.

Chapter 7 Answers

1. **a.** 100 mA × 0.30 sec = 30 mAs, the highest density of the choices.
2. **c.** Grids should be utilized with large-sized patients and when kVp is higher than 80, but patient dose will increase.
3. **b.** 4 × (3/1) = 12 mAs
4. **b.** 1.3 = 15.2/x, x = 11.7 cm
5. **b.** Osteoporosis is a destructive condition which requires the use of low exposures.
6. **d.** High ratio grids absorb more scatter and require more technique due to the absorption of some of the useful primary beam as well.
7. **d.** The transport system uses rollers to move the film through the processor and change its direction at the end of the chemistry wells.
8. **a.** Increasing the kVp leads to a longer scale of contrast.
9. **d.** mAs is not related to recorded detail.
10. **c.** The law of reciprocity describes the relationship between mAs, mA, and time.
11. **c.** Proper spectral matching between the color of the light emitted by the phosphor and the color of the light the film is sensitive to is the most important characteristic of screen/film.
12. **a.** Increased SID is necessary for increased OID.
13. **c.** The shortest time possible should be used if motion is a concern.
14. **a.** The anode heel effect describes more photons on the cathode side of the beam.
15. **d.** mA, time, and kVp all influence density.
16. **c.** The quantity of photons refers to mAs.
17. **c.** A compensating filter should be used on body parts with unequal thickness.
18. **a.** The lowest kVp produces the shortest scale of contrast.
19. **a.** Grid ratio = height of lead strips compared to interspace distance.
20. **b.** (18/x) = (422/382), x = 15 mAs
21. **c.** This is the definition of radiographic contrast.
22. **c.** Artifacts are any unwanted densities present on a film post-processing.
23. **d.** The higher the grid ratio, the better the scatter cleanup.
24. **c.** Contrast's primary control is kVp.
25. **d.** Added plus inherent equals total filtration.
26. **b.** Collimators do not clean up scatter after its been produced in the patient.
27. **c.** Mishandling of films can lead to pick-off of emulsion which appears as a density.
28. **d.** Due to the shortest distance, this set of technical factors would produce the highest density.
29. **b.** The developer chemistry will act to change the invisible manifest image into the visible latent image.
30. **c.** mAs does not contribute to contrast.
31. **b.** When high ratio grids are utilized, higher exposures must be used, thereby increasing patient dose.
32. **b.** The toe portion represents underexposure.
33. **d.** High kVp, low mA, and fast screen speeds all contribute to quantum mottle.
34. **a.** The lowest kVp produces the shortest scale of contrast.
35. **a.** Magnification and foreshortening are types of size distortion.
36. **c.** SID, OID, and focal spot size all affect recorded detail.
37. **a.** Density is the overall blackening on a film.
38. **d.** Digital radiography allows for wider technical latitude.
39. **c.** This set of exposure factors has the highest kVp and no use of a grid, which equals the greatest density.
40. **b.** x = 40/(40 − 4), x = 1.1
41. **d.** For every +1 there is an increase of 25% of density.
42. **a.** By blurring the grid strips, the image will appear with a limited grid line artifact.
43. **a.** The lowest grid ratio would present the longest scale of contrast.

44. a. 1 cm changes measured with a caliper varies the kV by 2.

45. c. Under humid conditions, pulling the film too quickly from a film bin leads to static.

46. b. A phosphor emits light after stimulation by a radiation source.

47. c. Elongation is the type of distortion that occurs when the primary beam is angled against the long axis of the part.

48. d. kVp relates to a beam's peak energy level, its penetrating power, and its quality.

49. b. Compensating filtration allows for a higher contrast image by evening out the densities.

50. a. Grids help assist in increasing an image's contrast.

51. c. The thicker portions of a patient's anatomy should be positioned toward the cathode end of the tube to best use the anode heel effect.

52. c. As described by the inverse square law, doubling the distance reduces the mAs by a factor of 4.

53. b. The latent image is the invisible image.

54. c. Photostimulable phosphors are used in CR.

55. b. The acceptable range of OD is 0.25 – 2.0.

56. d. Speed, latitude, and contrast can all be discerned from the characteristic curve.

57. b. The replenishment system activates at every 14 inches of film sent into the processor.

58. c. $x = 36/(36 - 3)$, $x = 1.1$.

59. a. $x = 10$ mAs $\times (200/400)$, $x = 5$.

60. b. Slower speed film requires more exposure in order to be properly exposed, causing density to decrease.

61. c. An upside down grid presents as severe cutoff along the periphery of the film.

62. a. Large grain film construction is low contrast film.

63. a. The shoulder portion represents overexposure.

64. c. Low speed screens require higher exposures to acquire adequate density, and they provide a greater visibility of detail.

65. a. Contrast will increase due to the restriction of the beam, which limits scatter production.

66. d. The replenishment system is responsible for the introduction of fresh chemistry into a processor.

67. c. Only kVp and mA can be adjusted by a technologist when using an AEC.

68. b. If not properly centered over an AEC's ionization chamber, the exposure will terminate too soon and the part will appear underpenetrated.

69. b. Quantum mottle appears as graininess on a high speed film when low mA and high kV are used.

70. b. Empyema, pus in the pleural cavity, is an additive pathology and requires a decrease in technology.

71. d. A minimum of 100 kVp should be used when doing barium contrast exposures.

72. a. A wedge filter should be used for a foot where the thicker portion can be placed under the thinnest part of the anatomy.

73. c. Densitometers are used to measure density.

74. b. The emulsion is made up of a combination of gelatin and silver halide crystals.

75. c. Film should be stored at temperatures below 68 degrees.

76. b. An artifact is an unwanted density on a processed film.

77. c. Both a grid and the air gap technique are used as means of reducing scatter from degrading a film.

78. b. Filtration is the removal of low energy photons.

79. b. Recorded detail is the degree of sharpness of a singular object.

80. a. Adjusting the window width allows more of the gray tones to stand out.

81. c. The focal spot size will not affect the amount of dosage a patient receives.

82. a. Size of the film controls the replenishment system.

83. c. kVp represents the penetrating power of the beam.

84. b. Increases in the OID will lead to magnification, which decreases recorded detail.

85. a. Rollers serve to transport film through an automatic processor.

86. b. An increase in photon energy (high kVp) would lead to an overall decrease in contrast.

87. a. By increasing the mAs, density is increased proportionally.

88. b. The film will be exposed; however, the film will appear with a decrease in contrast along the entire image.

89. b. This describes the 15% rule.

90. b. Only angling the tube to the plane of the IR will lead to elongation.

91. c. Slow speed films present better detail due to their small grain; the smaller the focal spot, the finer the detail of the image.

92. a. $x = 20 \times (200/800)$, $x = 5$

93. a. Short scale refers to a film having few densities that appear very different.

94. c. The wire mesh test is used to test screen-film contact.

95. d. Crossover rollers serve to move a film from one well to the next and are found out of solution.

96. a. Increasing kVp leads to decreases in contrast.

97. b. $100 \text{ mAs} = x \times 0.25 \text{ sec}$, $x = 400 \text{ mA}$

98. d. $800 \text{ mA} \times 0.05 \text{ sec} = x$, $x = 40 \text{ mAs}$

99. c. A focused grid's strips are angled to match the divergence of the beam.

100. b. This is the definition of spatial resolution.

RADIOGRAPHIC PROCEDURES AND POSITIONING

CHAPTER SUMMARY

Radiographic procedures are the backbone of diagnostic radiography. Utilizing the required positions, along with basic knowledge of a patient's anatomy and physiology, the radiographer will be able to produce images so physicians can provide a quick and accurate diagnosis.

Radiographic procedures consist of 60 questions, or 30% of the ARRT licensing examination. Basic gross anatomy and physiology, with emphasis on the skeletal system, is knowledge the radiographer needs in order to provide the proper radiograph projections. The protocols for each anatomical section differ in terms of positioning, and the radiographer must have the skills to modify protocols when required.

Key Terms

acromion process	ampulla of vater	barium sulfate
anterior	anatomical planes	bicipital groove
angle	anatomical position	bifid spinous process
ASIS	ankle joint	biliary tract
axial	anus	body habitus
acanthiomeatal line	articulating facets	brachycephalic skull
acanthion	articulating processes	bregma
acetabulum	asthenic	bronchi
acromioclavicular joint	astragalus	bronchus
adduction	atlas	base of fifth metatarsal
adrenal glands	auricle	calcaneous
ala	axillary border	catheter
alveoli	axis	cortex

Key Terms *continued*

carina	colon	cystourethrography
carpal bones	common bile duct	descending colon
caudad	condyle	degrees
cavities	contrast media	dens
cecum	coronal plane	duodenum
central ray	coronoid process	diaphragm
cephalad	costal cartilage	distal
cervical	cranium	dolichocephalic skull
chambers	cricoid cartilage	dorsal decubitus
cholangiography	cuboid	dorsiflexion
chyme	cuneiform	duodenum
clavicle	cystic duct	diaphragm
coccyx	cystograms	digestion

Key Terms *continued*

endocrine	fingers	gonion
ERCP	fissures	greater tubercle
enzymes	flat bone	gladiolus
epicondyles	flexion	glenoid fossa
erect	foot	hallux
esophagus	forehead	heel
eversion	fowlers	hepatic ducts
exhalation	frontal bone	hepatic flexure
extension	foramen magnum	hip
external acoustic meatus	fovea capitis	hip joint
false ribs	frontal sinuses	horizontal
femoral head	glabella	horizontal plane
femur	glabelloalveolar line	humerus
fibula	glenoid	hyoid bone

Key Terms *continued*

hyperextension	internal urethral orifice	kyphotic curvature
hyperflexion	interpupillary line	lambda
hyposthenic	intertrochanteric crest	lamboidal suture
ileocecal valve	intertrochanteric line	lamina
iliac crest	intervertebral discs	large intestine
ilium	intravenous urography	larynx
inferior	inversion	lateral
infraorbital margin	ischial tuberosity	lateral decubitus
infraorbital meatal line	ischium	left anterior oblique
infraspinatus fossa	insulin	left posterior oblique
inhalation	intercondyloid eminence	ligamentteres
inion	intercondyloid fossa	linea aspera
inner cantus	interphalangeal joints	liver
innominate bones	jugular notch	lobes

Key Terms *continued*

long bone	metacarpals	odontoid process
lordotic curve	metacarpophalangeal joint	olecranon fossa
lower extremity	metatarsals	olecranon process
lumbar	micturition	opacification
lunate	midaxillary	orbital meatal line
lungs	midcoronal plane	os calcis
malleolus	midsagittal	os magnum
manubrium	midsagittal plane	outer cantus
medial	minor calyx	overflexion
median planes	mortise	parallel
medulla	nasion	parietal bones
mental point	navicular	pedicles
mentomeatal line	oblique plane	perpendicular
mesocephalic skull	occipital bone	pelvic cavity

Key Terms *continued*		
pelvis	radial notch	rotator cuff
peristalsis	radial tuberosity	rugae
peritoneum	radius	sacroiliac joints
phalanges	rectum	shoulder joint
pisiform	red blood cells	sigmoid colon
posterior	renal	sinus tarsi
posterior superior iliac spine	renal pelvis	skull
post-void	retrograde pyelography	small bowel studies
pronation	retroperitoneal structures	small intestine
prone	ribs	spinal cord
proximal	right anterior oblique	spine
PSIS	right lateral recumbent	spinous process
pubis	right posterior oblique	spleen
pylorus	rotation	splenic flexure

Key Terms *continued*

squamosal suture	tailbone	tilt
sternal angle	talus	trachea
sternoclavicular joints	tangential	transverse
sternum	tarsals	transverse colon
sthenic	temporal bone	transverse process
stomach	terminal ilium	trapezium
styloid process	thoracic	trapezoid
subtalar joint	thoracic aorta	Trendelenburg
superior border	thoracic cavity	trigone
supine	thoracic spine	triquetrum
suprarenal	thoracic vertebrae	trochanters
supraspinatus fossa	thorax	trochlea
surgical neck	tibia	trochlear notch
sutures	tibial spine	true ribs

Key Terms *continued*		
t-tube	urinary system	vertebral column
tuberosity	ventral decubitus	vertex
urinary bladder	ventricles	villi
ureteral orifices	vertebrae	wrist
ureters	vertebral arch	wrist joint
urethra	vertebral body	xiphoid process
urinary bladder	vertebral border	zygapophyseal joints

Concepts and Skills

Radiographic procedures are an essential tool in the treatment of patients. In this chapter, the sections are broken down into the various body parts with their related organ systems, where applicable. Related radiographic positions follow each anatomical section. The review of procedures and positioning is organized as follows:

- classification of body habitus
- anatomical planes, directional movement, and positioning terminology
- the thorax and respiratory system
- the abdominal cavity, including the gastrointestinal system and genitourinary systems
- the spinal column
- both upper and lower extremities, including the shoulder girdle and pelvis
- cranium and facial bones

Classification of Body Habitus

Body habitus is the medical term for the physique or body type of a patient. There are four major categories for classifying patient body types. The four categories are as follows:

1. **Hypersthenic:** Approximately 5% of the population is of this body type. Hypersthenics have massive builds with a deep and broad thorax. The diaphragm is elevated. The stomach and gallbladder are very high and almost horizontal.
2. **Sthenic:** Approximately 50% of the population are sthenics. They have an athletic build, and the diaphragm is not as high. The thorax is slightly longer and narrow.

3. **Hyposthenic:** Approximately 35% of the population are hyposthenics. They are slender and light in weight with the stomach and gallbladder situated high in the abdomen.
4. **Asthenic:** Approximately 10% of the population are asthenics. They are extremely thin with a narrow and shallow thorax. The diaphragm, gallbladder, and stomach are situated low in the abdomen.

Anatomical Planes, Directional Movement, and Positioning Terminology

Anatomical directional terms can also be applied to the planes of the body. Body planes are used to describe specific regions of the body. These terms are used frequently and must be studied thoroughly.

Coronal (frontal) plane: Any plane dividing the patient into anterior (front) and posterior (back) sections. The midcoronal plane divides the patient into equal anterior and posterior portions. The midaxillary plane is a coronal plane that passes through the axilla at the junction of the arm and body when the arm is at a 90 degree angle with the body.

Sagittal plane: Any plane dividing the patient into right and left portions. The midsagittal and median planes divide the patient into equal right and left halves.

Transverse/horizontal plane: Any plane passing through the patient at a right angle to the coronal or sagittal planes. This also divides the patient into superior and inferior sections.

Oblique plane: Any plane that is not identified as a coronal, sagittal, or transverse/horizontal plane. This is primarily referenced in MRI and ultrasound contexts.

Anterior: Front portion of the body.

Posterior or ventral: Back portion (posterior) of the body.

Medial: Toward the medial plane or middle of the body.

Lateral: Opposite the medial plane, away from the middle of the body.

Proximal: Body part closest to the point of origin.

Distal: Opposite proximal; part farthest from the point of origin.

Cephalad: Toward the head.

Caudad: Toward the tail or feet.

Movement Terminology

The following directional terms are used to describe movement pertaining to the limbs. These terms are frequently used by technologists in day-to-day operations, and professionals must be perfectly fluent in their use.

Flexion: Decrease in the angle of the joint by the movement of bending. The term *hyperflexion* is the same as overflexion.

Extension: Increasing the angle of the joint by extending or straightening the joint. The term *hyperextension* is extending the joint beyond its normal limitations.

Dorsiflexion: Flexion between the lower leg and the foot, ultimately decreasing the angle between the two body parts to less than 90 degrees.

Inversion: Turning the foot inward at the ankle joint.

Eversion: Turning the foot outward at the ankle joint.

Abduction: Movement of body part away from the midline.

Adduction: Movement of the body part toward the midline.

Pronation: Turning of the hand so the palm is down.

Supination: Turning of the hand so the palm is facing upward.

Rotation: Circular motion around a specific axis.

Tilt: Movement in which the sagittal plane is not parallel to the long axis of the body part or table.

Positioning Terminology

The following terms best describe the general body positions commonly used in the field of radiography. Anatomical positions are to be used in conjunction with the anatomical planes, body habitus, and movement of patient.

Anatomical position: Standing erect with palms facing outward and all anterior surfaces of the body facing forward.

Supine: Lying down flat on your back, facing up.

Prone: Lying face down.

Right lateral recumbent (RLR): Erect or recumbent position, 90 degrees from true AP or PA.

Right posterior oblique (RPO): Right posterior side of patient closet to the table.

Left posterior oblique (LPO): Left posterior side closest to the table.

Right anterior oblique (RAO): Right anterior side of the body closest to the film.

Left anterior oblique (LAO): Left anterior side of the body closest to the table.

Dorsal decubitus: Patient supine with central ray passing horizontally.

Ventral decubitus: Patient is prone with central ray passing horizontally.

Lateral decubitus: Patient lies on either left or right side with the central ray passing either AP or PA.

Trendelenburg: Patient is supine with head lower than feet.

Fowlers: Head elevated 45–90 degrees.

Axial: Angling of the central ray greater than 10 degrees.

Tangential: To skim the surface of a part.

Skull Landmarks

Skull landmarks are standard reference points in the face and head used by technologists to determine angle, position, and plane.

Midsagittal plane: Separates the head into two symmetrical halves when viewed anteriorly.

Interpupillary line: This line connects the centers of the orbits and is at a 90 degree angle to the median sagittal plane.

Outer cantus: The most lateral, outer portion of the eye (almond).

Inner cantus: The most medial, inner portion of the eye (almond).

Acanthion: Also known as the anterior nasal spine, this is the area between maxilla and bottom of nose.

Gonion: Angle of mandible.

Nasion: Intersection of the frontal and two nasal bones, located just superior to the bridge of the nose.

Inion: The occipital protuberance.

Glabella: Located in the space between the eyebrows, it merges the two superciliary ridges.

Mental point: The most anterior point of the mandible in the midline and the most prominent point on the chin.

Infraorbital margin: Inferior portion of the orbit.

External acoustic meatus (EAM): The canal within the ear between the ear flap and the eardrum.

Auricle: Exterior cartilage portion of the ear.

Glabelloalveolar line: Passes through the anterior portion of the glabella, nasion, and the acanthion when in a lateral position.

Glabellomeatal line: Runs from the glabella to the EAM.

Orbital meatal line: The original "baseline" which runs from the nasion through the outer canthus of the eye to the center of the external auditory meatus.

Infraorbital meatal line: Slightly inferior to the orbital meatal line, it runs from the infraorbital margin to the EAM.

Mentomeatal line: Runs from the mental point to the EAM.

Acanthiomeatal line: Runs from the acanthion to the EAM.

Cranium Side View

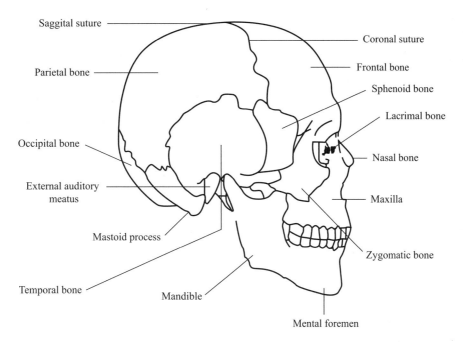

Saggital suture

Parietal bone

Occipital bone

External auditory meatus

Mastoid process

Temporal bone

Mandible

Mental foremen

Coronal suture

Frontal bone

Sphenoid bone

Lacrimal bone

Nasal bone

Maxilla

Zygomatic bone

Facial Bones

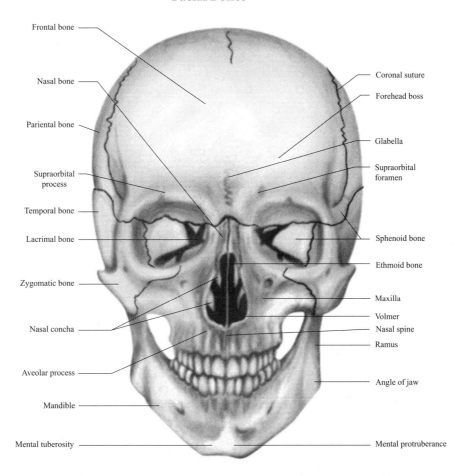

Frontal bone

Nasal bone

Pariental bone

Supraorbital process

Temporal bone

Lacrimal bone

Zygomatic bone

Nasal concha

Aveolar process

Mandible

Mental tuberosity

Coronal suture

Forehead boss

Glabella

Supraorbital foramen

Sphenoid bone

Ethmoid bone

Maxilla

Volmer

Nasal spine

Ramus

Angle of jaw

Mental protruberance

Shapes of the Skull

There are three major classifications of the skull. The following shapes of the skull reference the measurement of the angle between the petrous pyramids and the midsagittal plane to determine their classification.

Mesocephalic skull: Typical skull. The petrous pyramids project anteriorly and medially at an angle of 47 degrees from the midsagittal plane (MSP).

Brachycephalic skull: Short from front to back, broad from side to side, and shallow from vertex to base. The petrous pyramids are slightly wider from the MSP and form an angle of 54 degrees.

Dolichocephalic skull: Long from front to back, narrow from side to side, and deep from vertex to base. The petrous pyramids form a narrower angle with the MSP to create an angle of 40 degrees.

Thorax and Respiratory System
Thorax, Including Lungs

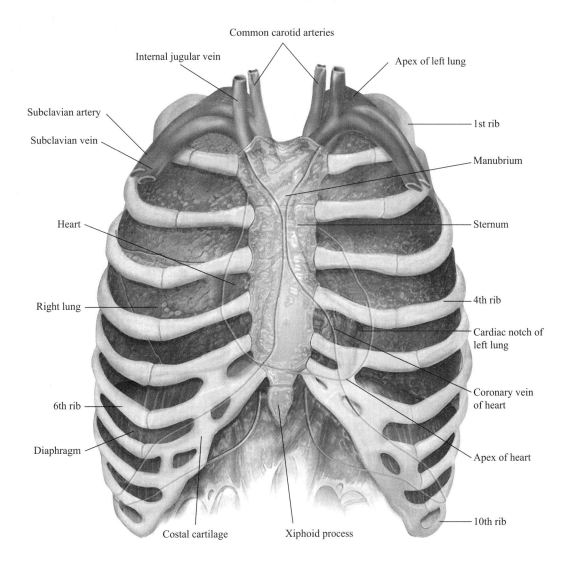

The thorax (also called the chest) is the region of the body formed by the sternum, the thoracic spine (vertebrae), and the ribs, which provide protection to the inner organs and are commonly referred to as the bony thorax. The region extends from the neck to the diaphragm. The heart, lungs, and esophagus reside in the thoracic cavity.

Sternum

The sternum, commonly referred to as the breastbone, is made up of three parts, and its margins articulate with the cartilages of the first seven pairs of ribs. The most superior portion is the manubrium, which contains a slight depression on its superior border called the jugular notch. Each clavicle articulates with the manubrium at the sternoclavicular joints. The midsection of the sternum is called the body, or gladiolus. The junction between the manubrium and the body is known as the sternal angle. The third inferior segment is known as the xiphoid process and is easily palpable.

Ribs

The ribcage is made up of 24 ribs; these ribs are divided up into 12 pairs, which articulate with the 12 thoracic vertebrae posteriorly in two places. The tuberosity of the rib articulates with the transverse process of the individual vertebra, while the head of the rib articulates directly with the body of the vertebra. Individually, they are long, curved bones. The first seven pairs of ribs, also referred to as true ribs, are directly attached to the sternum anteriorly through costal cartilage. The lower five pairs of ribs are designated as false ribs and are indirectly attached to the sternum by cartilage.

Larynx

The larynx (voice box) assists in protecting the trachea. It is found between the levels of C-3 and C-6 vertebrae and extends vertically from the tip of the epiglottis to the inferior border of the cricoid cartilage. Although the hyoid bone is not considered part of the larynx, it is located anteriorly at the level of C-3.

Trachea

The trachea is also commonly referred to as the windpipe and allows for the passage of air toward the lungs. It is located anteriorly to the esophagus in the mediastinum and extends from C-6 to T-5, where it bifurcates into the right and left main bronchus at carina.

Bronchi

The right and left bronchi transport air into the lungs. The right main bronchus is wider, shorter, and more vertical than the left and is of importance during the insertion of endotracheal tubes.

Lungs

The lungs are the organs of respiration in the body. They serve to transport oxygen into the bloodstream as well as to remove carbon dioxide waste from the body. This exchange takes place in the lungs at the alveoli. Both lungs are separated into lobes by fissures. There are three lobes in the right lung and two lobes on the left.

Heart

The heart is a muscular organ responsible for pumping blood throughout the body by continuous contractions and relaxations. It is offset slightly to the left of the midline thorax. It has four chambers, two superior atria and two inferior ventricles. The right side of the heart is responsible for transporting deoxygenated blood into the lungs, while the oxygenated blood is circulated through the body via the left side.

Thoracic Aorta

The aorta transports oxygenated blood and arises from the left ventricle of the heart where it proceeds upward until the level is at T-3 to T-4. It then arches downward (aortic arch) posteriorly midline through the thoracic cavity, then descends (descending aorta).

Diaphragm

The diaphragm is a dome-shaped, flat muscle that separates the thoracic and abdominal cavities. This muscle contracts during inhalation, thus increasing lung capacity, and retracts during exhalation.

Positions

Note: All positions covered in this chapter require gonadal shielding on all patients regardless of reproductive age. As long as the shield will not in any way interfere with the diagnosis of a patient (e.g., exams of the abdomen in females), shielding should be used.

	Position of Patient	Position of Part	Central Ray	Structures
PA Chest	Erect, shoulders rotated forward to limit scapulae in lung field	MSP of body to ML of wall unit; exposure at second full inspiration	SID 72 inches perpendicular to MSP at level of T-7 (inferior angle of scapula)	Lungs (apices, costophrenic angles, lateral margins), heart with magnification limited due to SID, trachea, 10 posterior ribs
Lateral Chest	Erect, left side against wall unit; arms raised above head	Metacarpalphalangeal (MCP) of body to Mid Line of wall unit; exposure at full inspiration	SID 72 inches perpendicular to MCP at level of T-7	Superimposed lungs (apices, costophrenic angle), heart, bony thorax
PA Oblique Chest (RAO/LAO)	Erect, side of interest closest to wall unit; opposite arm elevated	Patient rotated 45 degrees away from affected side, exposure at inspiration	SID 72 inches perpendicular at level of T-7	LAO—right lung, RAO—left lung; for dedicated heart study, rotate patient 60 degrees
Lateral Decubitus Chest	Lateral recumbent; position elevated on board	Center chest to IR; exposure at inspiration	SID 72 inches horizontal and perpendicular to the area of T-7	Entire lung fields; air levels—affected side up; fluid levels—affected side down
Lordotic Chest	Patient AP erect 1 foot in front of wall unit (if patient cannot tolerate erect position, can be done supine)	MSP of body to ML of wall unit; have patient lean back until shoulder touches wall unit	SID 72 inches (if supine maximum SID allowable) horizontal to the ML entering mid-sternum; however, if done supine, angle CR 20 degrees cephalic, exposure at inspiration	Clavicles projected over the apices of the lungs allowing for clear visualization of both apices
AP Soft Tissue Neck (Airway—Upper)	Patient erect, however can be done supine	MSP of body to ML of wall unit, extend chin upwards; exposure during slow inhalation	40 inches SID perpendicular to the area of the thyroid cartilage	Region from C-3–T-3, air-filled pharynx, larynx, and trachea

	Position of Patient	Position of Part	Central Ray	Structures
Lateral Soft Tissue Neck (Airway—Upper)	Patient in upright lateral position	Center coronal plane which passes through the center of the soft tissue of the neck to the wall unit with slight extension of the chin; exposure during slow inhalation	72 inches SID perpendicular to a point 2 inches superior to the jugular notch at the ML of the IR	Region from C-1–T-3, air-filled pharynx, larynx, and trachea
AP Ribs (Above Diaphragm)	Patient in either upright or supine position; MSP of body to ML of unit for a bilateral exam or a sagittal plane of the body centered between the lateral border of the thorax and the spine	Attempt should be made to manipulate patient's arms so that scapula is out of the lung field; IR 2 inches above the shoulders; exposure during full inspiration	Perpendicular to the midpoint of the cassette	Posterior ribs best visualized above diaphragm
PA Ribs (Above Diaphragm)	Patient in either upright position facing wall unit, MSP of body to ML of unit for a bilateral exam, or a sagittal plane of the body centered between the lateral border of the thorax and the spine	Attempt should be made to manipulate patient's arms so that scapula is out of the lung field; IR 2 inches above the shoulders; exposure during full inspiration	Perpendicular to the midpoint of the cassette	Anterior ribs best visualized above the diaphragm
AP Ribs (Below Diaphragm)	Patient in either upright or supine position, MSP of body to ML of unit for a bilateral exam	Lower border of IR 1 inch below the level of iliac crest; exposure during full exhalation	Perpendicular to the midpoint of the cassette	Ribs below the diaphragm best visualized utilizing the contrast of the upper abdomen
AP Oblique Ribs (RPO/LPO)	Patient either erect or supine with affected region of interest closest to ML of unit	Patient rotated 45 degrees with affected side closest to IR; upper border of IR should be 2 inches above shoulder	Perpendicular to the midpoint of the cassette	Axillary ribs free of superimposition of the body

	Position of Patient	Position of Part	Central Ray	Structures
PA Oblique Ribs (RAO/LAO)	Patient either erect or prone with affected region of interest furthest centered to ML of unit	Patient rotated 45 degrees with affected side furthest from the IR; upper border of IR should be 2 inches above shoulder	Perpendicular to the midpoint of the cassette	Axillary ribs free of superimposition of the body
RAO Sternum	Patient in semi-prone position	Patient rotated in RAO position 15–20 degrees; upper border of IR to the level of C-7; exposure to be done during full expiration or utilizing breathing technique to enhance contrast	Perpendicular to the midpoint of the sternum	Sternum visualized through the shadow of the heart
Lateral Sternum	Patient in upright lateral position	Center sternum to the ML of the unit; upper border of IR approximately 1.5 inches above the sternal notch	72 inches SID perpendicular to the midpoint of the sternum	Entire sternum visualized
PA Sternoclavicular Joints	Patient in prone or seated position	Center MSP to ML of the unit; center IR to the jugular notch (level of T-3)	Perpendicular to the midpoint of the cassette entering T-3	Bilateral SC joints visualized and medial clavicles
PA Oblique Sternoclavicular Joints	Patient in prone or seated position	Affected side closest to IR; rotate patient approximately 10–15 degrees with joint centered to ML of grid	Perpendicular to the SC joint closest to the IR entering at the level of T-2–T-3	

Abdomen and Gastrointestinal System

Digestive System

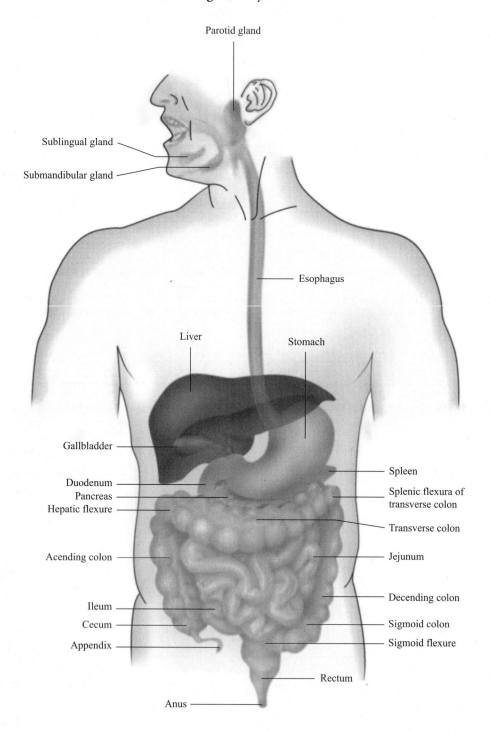

Parotid gland

Sublingual gland

Submandibular gland

Esophagus

Liver

Stomach

Gallbladder

Spleen

Duodenum

Splenic flexura of transverse colon

Pancreas

Hepatic flexure

Transverse colon

Acending colon

Jejunum

Decending colon

Ileum

Cecum

Sigmoid colon

Appendix

Sigmoid flexure

Rectum

Anus

The abdominal cavity is located between the diaphragm superiorly and an imaginary line located at the level of the iliac crest of the pelvis. It is continuous with the pelvic cavity below, which contains some of the organs of the gastrointestinal tract as well as the reproductive organs. The peritoneum is the serous membrane that surrounds the various organs. Organs of the abdominal cavity include the majority of the large intestine, the stomach, small intestine, liver, gallbladder, pancreas, spleen, kidneys, and ureters. Organs of the pelvic cavity include the sigmoid colon, rectum, and urinary bladder.

Esophagus

The esophagus is a hollow muscular tube that passes food from the pharynx via muscle contractions (peristalsis). Peristalsis takes place throughout the entire digestive system and is strongest at the esophagus.

Stomach

The stomach is located between the esophagus and small intestine and is the primary organ of digestion where food is converted into chyme by digestive enzymes. The stomach and esophagus meet at the cardiac orifice. When the stomach is empty, the mucosal lining falls into longitudinal folds called rugae. There are three portions of the stomach: the fundus (which is the most superior rounded portion), the body, and the pylorus.

Small Intestine

The majority of the absorption of nutrients takes place in the small intestine. It averages 20 feet in length and consists of three segments. The first segment is the duodenum, which is 10 inches in length and is usually C-shaped. The second segment is the jejunum, which measures 8 feet in length. The third and longest segment of the small intestine is the ilium, which measures 11 feet in length. The small intestine connects with the large intestine at the ileocecal valve. Materials are assisted in moving through the small intestine via villi.

Large Intestine

The large intestine is responsible for absorbing water and transiting waste out of the body. It is divided into three primary sections (cecum, colon, and rectum) and measures around 5 feet long in total. It is much wider than the small intestine and is differentiated by the appearance of haustra pockets. The cecum is the first aspect of the large intestine and is located at the right iliac region of the pelvis. The cecum extends upward into the ascending colon to a bend at the hepatic flexure located below the liver on the right. The transverse colon continues from this bend on a slightly upward plane until it reaches the splenic flexure. At this flexure, the colon turns downward (descending colon) until it reaches the left iliac area where it curves on itself within the pelvis. This area is termed the sigmoid colon. When the sigmoid reaches the midline, it turns sharply downward into the rectum. The rectum terminates at the anus.

Liver

The liver is located in the right upper quadrant of the abdomen just below the diaphragm. It is responsible for the formation of bile which assists in breaking down digested materials in the small intestine. It consists of four lobes. Bile is transported out of the liver through the right and left hepatic ducts at first. These two ducts eventually meet and merge at the common hepatic duct. The common hepatic duct later merges with the cystic duct from the gallbladder to form the common bile duct. The common bile duct terminates at the duodenum via the ampulla of vater.

Gallbladder

The gallbladder is a pear-shaped sac located at the inferior aspect of the liver. Its primary function is to store bile, which is transported from the gallbladder via the cystic duct.

Pancreas

The head of the pancreas is housed within the loop of the duodenum. The tail is located toward the spleen. It serves purposes in both the digestive and the endocrine systems. Insulin is produced in the pancreas.

Spleen

The spleen is located in the left upper quadrant of the abdomen and plays an important role in the formation and storage of red blood cells.

Preparation for Contrast Studies

It is important to note the special preparations that are necessary for imaging of the abdomen, especially when pertaining to contrast studies. For examinations of the upper gastrointestinal system, patients should be NPO (no food or drink) 8–9 hours prior to the procedure to ensure an empty digestive tract. Examinations of the lower gastrointestinal system are best performed when the colon is completely empty of any contents. This is generally achieved by having the patient consume a low-residue diet two days prior to the examination as well as cleansing the bowel using a laxative the day before. A scout film is normally obtained prior to contrast studies in order to make sure the proper preparation was followed. Generally, digestive contrast studies are performed as double contrast studies wherein air acts as a negative contrast medium in relation to the positive contrast media administered. Barium sulfate is the most common contrast medium for these studies unless contraindicated. Contraindications for barium include suspected perforation or severe intestinal obstruction. In these cases, a water-soluble contrast medium is used instead; however, imaging is limited due to the increased absorption of the medium as well as its dilution after ingestion. For enema studies, it is also important to note that to insert the enema tip the patient should assume the lateral Sims position. The tip of the catheter should be inserted into the anus no more than 3.5–4.0 inches. The contrast media bag should be located 18–24 inches above the level of the anus in order to optimize the initial flow of contrast into the colon.

Positions

	Position of Patient	Position of Part	Central Ray	Structures
AP Abdomen (KUB)	Patient in supine position, arms at side; center MSP to ML of grid device	Center IR to the level of iliac crest; bottom border of IR should be just distal to level of greater trochanter (symphysis pubis)	Perpendicular to the center of the IR	Abdomen including kidneys, liver, psoas muscle; symphysis pubis, sacrum, coccyx, lumbar spine, and distal thoracic spine
AP Abdomen— Upright	Patient in upright position, arms at side; center MSP to ML of upright grid device	Center IR to a level 2–3 inches above the iliac crests; upper border of IR should be above the level of the diaphragm	Perpendicular to the center of the IR	Abdomen including kidneys, liver, psoas muscle; diaphragm should be visualized as well as air/fluid level within the abdomen. Lumbar and distal thoracic spine

	Position of Patient	Position of Part	Central Ray	Structures
AP Abdomen—Left Lateral Decubitus	Patient lies in a lateral recumbent position with the left side of patient closest to the table; arms elevated above the patient's head; center MSP to ML of grid device	Center IR horizontal to ML of the unit; center IR to a level 2–3 inches above the iliac crest; if necessary, adjust IR to ensure the entire elevated side is included	Perpendicular to the center of the IR	Air/fluid levels within the abdomen, diaphragm
Abdomen Dorsal Decubitus	Patient lies in a supine position, arms elevated above head	Center IR to a level 2–3 inches above the iliac crest	Horizontal and perpendicular to the center of the IR	Air/Fluid levels within the abdomen, diaphragm
Esophagus—AP Projection	Patient lies in a supine position with arms at side; center MSP to ML of grid device; patient should swallow contrast during exposure	Center IR to a level 1 inch inferior to sternal notch; upper border of IR should be 2–3 inches superior to shoulders	Perpendicular to the center of the IR, approximately T-5–T-6	Contrast-filled esophagus and fundus of stomach
Esophagus—RAO Projection	Patient lies in a semi-prone position on table with the right side of the body toward the grid device	Patient rotated in an RAO position approximately 35–45 degrees; center IR to a level 1 inch inferior to sternal notch; upper border of IR should be 2–3 inches superior to shoulders	Perpendicular to the center of the IR, approximately T-5–T-6	Contrast-filled esophagus and fundus of stomach
Esophagus—LAO Projection	Patient lies in a semi-prone position on table with the left side of the body toward the grid device	Patient rotated in an LAO position approximately 35–45 degrees; center IR to a level 1 inch inferior to sternal notch; upper border of IR should be 2–3 inches superior to shoulders	Perpendicular to the center of the IR, approximately T-5–T-6	Contrast-filled esophagus and fundus of stomach

	Position of Patient	Position of Part	Central Ray	Structures
Esophagus— Left Lateral	Patient lies in a lateral recumbent position with the left side closest to the grid device; arms should be out front away from the body	Center IR to a level 1 inch inferior to sternal notch; upper border of IR should be 2–3 inches superior to shoulders	Perpendicular to the center of the IR, approximately T-5–T-6	Contrast-filled esophagus and fundus of stomach
Esophagus— PA Projection	Patient lies in a prone position with arms elevated; center MSP to ML of grid device; patient should swallow contrast during exposure	Center IR to a level 1 inch inferior to sternal notch; upper border of IR should be 2–3 inches superior to shoulders	Perpendicular to the center of the IR, approximately T-5–T-6	Contrast-filled esophagus and fundus of stomach
Upper GI Series— PA Stomach	Patient lies in a prone position with arms elevated	Center IR to a sagittal plane midway between the spine and the left lateral border of the abdomen at the level of L-2	Perpendicular to the center of the IR, at approximately L-2	Contrast-filled stomach and duodenal bulb; in asthenic or hyposthenic patients, the pyloric canal and duodenal bulb is best demonstrated
Upper GI Series— PA Oblique RAO Stomach	Patient lies in a semi-prone position on table with the right side of the body toward the grid device	Patient rotated in an RAO position approximately 40–70 degrees, depending on the size and shape of the stomach; hypersthenic patients require the steeper rotations; center IR to a sagittal plane midway between the spine and the lateral border of the elevated abdomen at the level of L-2	Perpendicular to the center of the IR, at approximately L-2	Contrast-filled stomach and duodenal bulb; in sthenic patients, the pyloric canal and duodenal bulb is best demonstrated; fundus filled with air; contrast opacified pylorus and duodenal bulb

	Position of Patient	Position of Part	Central Ray	Structures
Upper GI Series— Right Lateral	Patient lies in a lateral recumbent position with the right side closest to the grid device; arms should be out front away from the body	Center IR to a coronal plane midway between the MCP and the anterior surface of the abdomen at the level of L-2	Perpendicular to a coronal plane midway between the MCP and the anterior surface of the abdomen at the level of L-2	Contrast-filled anterior and posterior view of the stomach and duodenal bulb; in hypersthenic patients, the pyloric canal and duodenal bulb is best demonstrated
Upper GI Series—AP Oblique LPO	Patient lies in a semi-supine position on table with the left side of the body toward the grid device	Patient rotated in an LPO position approximately 30–60 degrees (45 degrees average), depending on the size and shape of the stomach. Center IR to a sagittal plane midway between the spine and the left lateral border of the depressed side of the abdomen at the level of L-2	Perpendicular to the center of the IR at the approximate level of L-2	Contrast-filled stomach and duodenal bulb; fundus of stomach opacified by contrast; air-filled pylorus and duodenal bulb

Small Bowel Studies

Small bowel studies are timed studies in which the patient ingests a positive contrast media, generally barium sulfate. Sequential KUB images are taken until the contrast reaches the cecum of the colon. Depending on the peristalsis of the small intestine and any pathology noted after ingestion, these studies can take upward of several hours. Upon reaching the cecum, the terminal ilium of the small bowel is viewed, with either fluoroscopy or a static image, by applying pressure onto the area just medial to the right ASIS utilizing a compression paddle apparatus with the patient in a right anterior oblique position.

Barium Enema— PA Projection	Patient in prone position, arms at side; center MSP to ML of grid device	IR centered to the level of the iliac crest	Perpendicular to the midline at the level of the iliac crest	Entire contrast-filled colon
Barium Enema— AP Oblique Projection (RPO/LPO)	Patient lies in a semi-supine position on table with the body side in question toward the grid device	Patient rotated in an AP oblique position approximately 35–45 degrees; IR centered to the level of the iliac crest	Perpendicular to the midline at the level of the iliac crest approximately 1–2 inches lateral to the ML of the body on the elevated side	RPO—best view of descending colon and splenic flexure; LPO—best view of ascending colon and hepatic flexure
Barium Enema— PA Oblique Projection (RAO/LAO)	Patient lies in a semi-prone position on table with the body side in question toward the grid device	Patient rotated in a PA oblique position approximately 35–45 degrees; IR centered to the level of the iliac crest	Perpendicular to the midline at the level of the iliac crest approximately 1–2 inches lateral to the ML of the body on the elevated side	RAO—best demonstrates ascending colon, sigmoid colon, and hepatic flexure; LAO—best demonstrates descending colon and splenic flexure
Barium Enema— AP Axial	Patient in supine position, arms at side; center MSP to ML of grid device	IR centered to a level 2 inches superior to the level of the iliac crest	30–40 degrees cephalic angle directed 2 inches inferior to the level of ASIS	Rectosigmoid region
Barium Enema— PA Axial	Patient in prone position, arms above head; center MSP to ML of grid device	IR centered to the level of the iliac crest	30–40 degrees caudal angle directed to the level of ASIS	Rectosigmoid region

Barium Enema— Left Lateral Rectum	Patient lies in a lateral recumbent position with the left side closest to the grid device; arms should be out front away from the body	Center IR to a plane 2 inches posterior to MCP to the midline of the grid device	Perpendicular to the level of ASIS	Rectum and sigmoid; distal descending colon
Barium Enema— Lateral Decubitus (Right/Left)	Patient lies in a lateral recumbent position with the side of interest closest to the table; arms elevated above the patient's head; center MSP to ML of grid device	Center the IR to the level of the iliac crest	Horizontal and perpendicular to enter the ML of the body at the level of iliac crest	Right lateral—contrast-filled colon, medial side of ascending colon, and lateral side of descending colon filled with air; left lateral—contrast-filled colon, medial side of descending colon and lateral side of ascending colon filled with air

Cholangiography

These procedures are performed for visualization of the biliary tract and the gallbladder and are most commonly performed in a sterile environment such as an operating room, utilizing mobile radiographic equipment. A small catheter is inserted into the common bile duct endoscopically, and contrast media is injected. This is called endoscopic retrograde cholangio pancreatography, or ERCP. Postoperative procedures require contrast media to be injected through a T-tube, which will have been inserted previously.

Urinary System

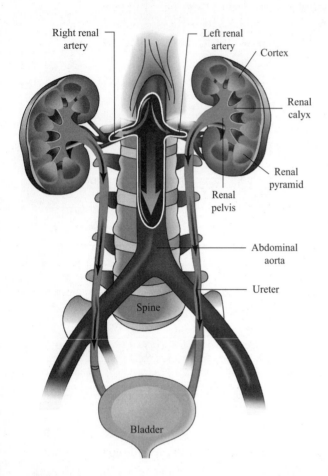

Right renal artery

Left renal artery

Cortex

Renal calyx

Renal pyramid

Renal pelvis

Abdominal aorta

Ureter

Spine

Bladder

Kidneys

The kidneys are bean-shaped organs, each approximately the size of a fist. They are located in the posterior abdomen and sit bilaterally at the levels of T-12 through L-3, with the right kidney slightly lower in position than the left. This is due to the large mass of the liver located in the right upper quadrant. The kidneys are also considered retroperitoneal structures due to their location behind the peritoneum. Their main function is to remove waste products from the blood and produce urine. The outer portion of the kidney is referred to as the cortex. The renal pyramids are the internal structures which end in a minor calyx. These minor calyces all unite to form the two or three major calyces, which in turn unite into the renal pelvis.

Ureters

The ureters are the pathway between the kidneys and the urinary bladder. Urine produced in the kidneys drains to the urinary bladder via the renal pelvis into the right and left ureters. They are hollow tubes measuring approximately 10 to 12 inches long and, like the kidneys, are located behind the peritoneum.

Urinary Bladder

The bladder serves as a main reservoir for urine produced by the kidneys. The base of the bladder rests on the pelvic floor. The trigone is a smooth triangular region formed by the two ureteral orifices and the internal urethral orifice. Micturition is the process of removing urine from the bladder via the urethra.

Urethra

The urethra serves as the passageway for urine to be excreted from the body. The urethra is the tube that connects the urinary bladder to the outside of the body.

Adrenal Glands

Adrenal glands are triangle-shaped glands located on top of the kidneys. They are also known as suprarenal glands due to their physical location. The outer part of the adrenal gland is called the cortex and the inner part of the adrenal gland is called the medulla. These structures are part of the endocrine system.

Procedures

Cystograms are procedures that are performed for visualization of the bladder. They require contrast media to be injected via catheter into the patient's bladder either through the urethra or through a surgically made opening. A water-soluble contrast media is preferred. Upon total opacification, the catheter is removed and the patient is required to withhold from urinating while images are performed. The positions for the bladder are as follows: AP projection, two AP oblique projections, and a lateral projection. During voiding cystourethrography (VCUG), exposures are made to visualize any possible anomaly that may be present.

Intravenous urography requires iodinated contrast media be injected into the bloodstream via IV catheter. Serial images of the kidneys, sometimes utilizing tomography, are sometimes taken as the contrast uptakes if the kidneys are of interest. This is then followed by AP and oblique abdomen images which must include the kidneys, ureters, and bladder for proper evaluation. Post-void images of the bladder are also taken. Retrograde pyelography takes place when a catheter is inserted into the urethra and guided directly into a ureter with contrast media then administered into the kidneys.

Vertebral Column

The vertebral column, or spine, is located posteriorly on the midsagittal plane of the body. The vertebral column consists of a group of 33 vertebrae and intervertebral discs. The entire column is broken up into groups: cervical, thoracic, lumbar, sacral, and coccyx. Each grouping has its own characteristic curvature. Each individual vertebra contains a number of similarities regardless of location.

Typical Vertebra Superior View

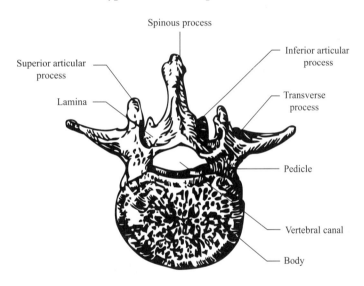

Typical Vertebra Lateral View

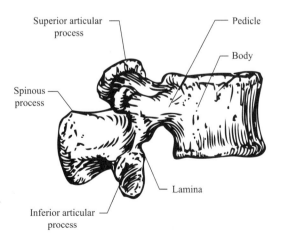

Vertebrae

The general construction of a typical vertebra consists of an anterior vertebral body. The body is the portion that bears weight and provides support. The posterior, ringlike portion is referred to as the vertebral arch. The vertebral arch encloses and protects the spinal cord. The arch is composed of two pedicles, one on either side. The pedicles support the laminae. The laminae join together at the midline to form the spinous process, which is palpable in both the thoracic and lumbar spines. On the left and the right of each vertebral arch extends a transverse process, which is for muscle attachment. On the superior and inferior portions of the arch, there are articulating processes, which form the zygapophyseal joints.

Cervical Spine

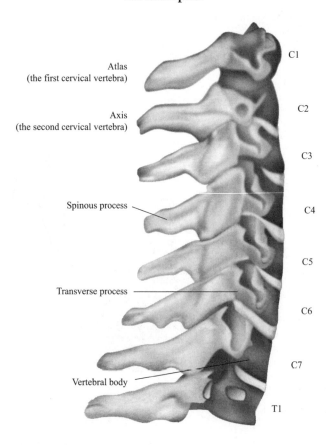

Atlas
(the first cervical vertebra)

Axis
(the second cervical vertebra)

Spinous process

Transverse process

Vertebral body

C1

C2

C3

C4

C5

C6

C7

T1

There are seven cervical vertebrae which together present a lordotic curve. Characteristics include:

- small vertebral body
- bifid spinous process
- C-1 (atlas) is ringlike and contains no body or spinous process. This provides support for the head and articulates with the skull to allow for flexion and extension.

- C-2 (axis) contains the odontoid process (dens) which rises superiorly from its body and articulates with the atlas above. This allows the head to rotate.
- C-7 contains a larger spinous process which is referred to as vertebra prominens. This is extremely pronounced and easily palpable.

- Each cervical vertebra contains a transverse foramen on the transverse processes which allows transit of the vertebral arteries upward into the brain.

Thoracic Spine

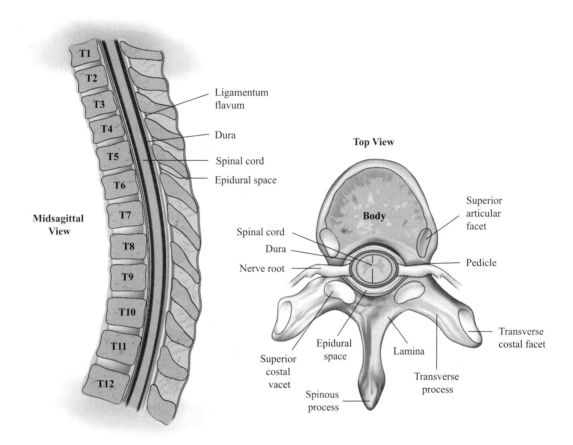

There are 12 thoracic vertebrae. They progressively increase in size inferiorly from the cervical region toward the lumbar region. They present a kyphotic curvature. Characteristics include:

- articulating facets on both the body and transverse processes for the ribs
- long and sharp inferior angles; spinous processes

Lumbar Spine

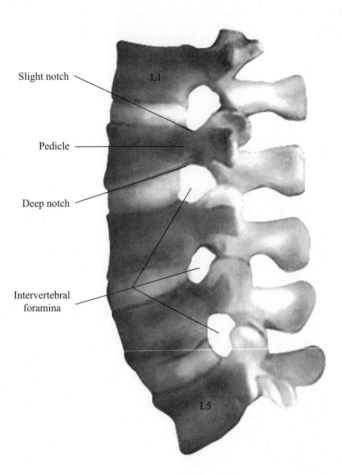

Slight notch

Pedicle

Deep notch

Intervertebral foramina

L1

L5

There are five lumbar vertebrae which present a lordotic curve. They are the largest and bear the most stress in their support of the body. The lumbar vertebrae take the appearance of a Scottie dog when placed at a 45 degree oblique angle. In this appearance, the apophyseal joints are illustrated where the ear of the Scottie dog represents the superior articulating process, the nose corresponds to the transverse process, the neck corresponds to the pars interarticularis, the front foot corresponds to the inferior articulating process, the eye corresponds to the pedicle, and the body corresponds to the lamina.

Sacrum

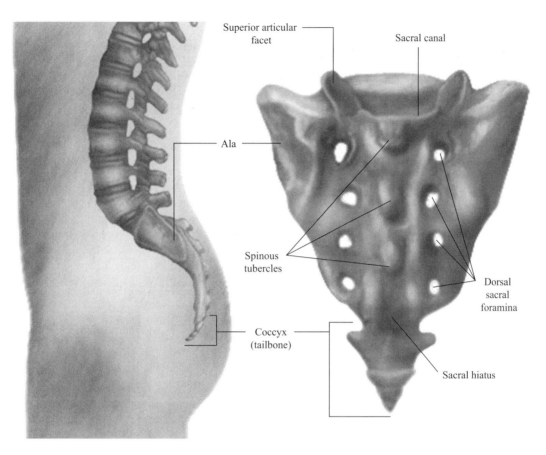

There are five sacral vertebrae which are fused. The transverse processes are referred to as the alae. The sacral foramina, located both anteriorly and posteriorly, transmit the spinal nerves. The sacrum articulates with the bony pelvis.

Coccyx
The coccyx consists of three to five fused vertebrae. It is commonly referred to as the tailbone.

Positions

	Position of Patient	Position of Part	Central Ray	Structures
Cervical Spine— AP Axial Projection	Patient is either erect or supine; MSP to ML of grid device	Extend chin so occlusal plane is perpendicular to grid device; center IR to level of C-4	15–20 degree cephalic angle directed through C-4	Lower 5 cervical vertebral bodies, upper thoracic spine
Cervical Spine— AP Open Mouth Projection	Patient is supine; MSP to ML of grid device; ensure no rotation of patient's head	Adjust patient's head so that a line drawn from the lower edge of the upper incisors to the mastoid tips is perpendicular to the IR; center IR to the level of C-2; ask patient to open mouth wide without moving head	Perpendicular through the patient's open mouth	Dens through the open mouth
Cervical Spine— Lateral Projection	Patient upright in a true lateral position (right or left); MCP of patient to ML of grid device; patient's shoulders should be depressed as much as possible Position can be obtained cross-table if the patient cannot assume upright position	Adjust IR so that it is centered at the level of C-4	SID 60–72 inches horizontal and perpendicular to C-4; cross-table effort must be made to obtain the greatest SID (up to 72 inches) possible	All seven vertebral bodies; cervical zygapophyseal joints
Cervical Spine— AP Oblique Projection (RPO/LPO)	Patient upright or semi-supine	Adjust rotation of body to obtain a 45 degree angle with the plane of the IR; IR should be centered at the level of C-3	SID 40 or 72 inches, 15–20 degree cephalic angle direct to C-4	Intervertebral foramina and pedicles farthest from the IR

	Position of Patient	Position of Part	Central Ray	Structures
Cervical Spine— PA Oblique Projection (RAO/LAO)	Patient upright or semi-prone	Adjust rotation of body to obtain a 45 degree angle with the plane of the IR; IR should be centered at the level of C-5	SID 40 or 72 inches, 15–20 degree caudal angle direct to C-4	Intervertebral foramina and pedicles closest to the IR
Cervicothoracic Spine—Lateral Projection (Swimmer's)	Patient erect in lateral position; MCP to ML of grid device; arm closest to IR should be elevated over head; arm farthest from IR should be depressed as best as possible	Adjust IR to center at the level of C-7–T-1	Horizontal and perpendicular to C-7; if the shoulder cannot be fully depressed, a 5 degree caudal angle can be utilized	Lower cervical and upper thoracic spine visualized
Cervical Spine— Lateral Flexion Projection	Patient upright in a true lateral position (right or left); MCP of patient to ML of grid device; patient's shoulders should be depressed as much as possible	Adjust IR so that it is centered at the level of C-4; depress chin as much as possible to obtain hyperflexion	SID 60–72 inches horizontal and perpendicular to C-4	Cervical vertebrae; spinous processes should be separated
Cervical Spine— Lateral Extension Projection	Patient upright in a true lateral position (right or left); MCP of patient to ML of grid device; patient's shoulders should be depressed as much as possible	Adjust IR so that it is centered at the level of C-4. Extend chin as much as possible to obtain hyperextension	SID 60–72 inches horizontal and perpendicular to C-4	Cervical vertebrae; spinous processes should be closer in proximity
Dens—AP Projection (Fuchs Method)	Patient supine; MSP to ML of grid device; ensure no rotation of head	IR at level of the mastoid tips; extend chin so that it lies on the same plane of the mastoid tips	Perpendicular to the IR just distal to the tip of the extended chin	Dens visualized through the foramen magnum

	Position of Patient	Position of Part	Central Ray	Structures
Dens—PA Projection (Judd Method)	Patient prone; MSP to ML of grid device; ensure no rotation of head	Extend patient's neck so that the chin rests on the table; IR centered to the level of the thyroid cartilage	Perpendicular to the IR just distal to the edge of the mastoid tips	Dens visualized through the foramen magnum
Thoracic Spine— AP Projection	Patient supine; MSP to ML of grid device	Center the upper margin of the IR approximately 1.5–2.0 inches superior to the shoulders	Perpendicular to the IR at the level of T-7 directed to the MSP	Thoracic bodies, intervertebral disc spaces
Thoracic Spine— Lateral Projection	Patient in lateral recumbent position; MCP to ML of grid device; arms extended in front of the patient	Center the upper margin of the IR approximately 1.5–2.0 inches superior to the relaxed shoulders; center the posterior half of the back to the IR	Perpendicular to the IR at the level of T-7; exam can be done on full expiration or utilizing a breathing technique in order to blur out overlying structures	Thoracic intervertebral foramina, spinous processes, thoracic bodies
Lumbar Spine— AP Projection (PA Projection Optional)	Patient supine, arms at side; MSP to ML of grid device; flex patient's knees and place feet flat on table to reduce the normal lordotic curvature of the lumbar spine. Can be done PA in order to reduce direct dose to the gonads	IR centered to the level of the iliac crest	SID 48 inches perpendicular to the IR at the level of iliac crest (L-4)	Five lumbar vertebrae
Lumbar Spine— Lateral Projection	Patient in lateral recumbent position with knees flexed; MCP to ML of grid device; laterality depends upon affected side	Adjust IR height to the level of the iliac crest	Perpendicular to the IR at the level of iliac crest (L-4)	Lumbar intervertebral foramina; disc spaces

	Position of Patient	Position of Part	Central Ray	Structures
L5–S1—Lateral Spot Projection	Patient in lateral recumbent position with knees flexed; MCP to ML of grid device; laterality depends upon affected side	Center IR on a coronal plane 2 inches posterior to ASIS and 1.5 inches inferior to iliac crest	Perpendicular to the IR with close collimation; if the spine is not horizontal, a 5 degree caudal angle can be utilized for male patients; an 8 degree caudal angle can be utilized for female patients	Open lumbosacral joint; tight collimation should be utilized for optimum contrast
Lumbar Spine— AP Oblique Projection (RPO/LPO)	Patient semi-supine; rotate patient 45 degrees toward the affected side	Center the rotated spine to the ML of the grid device located 2 inches medial to the elevated ASIS; adjust IR to a level 1.5 inches above iliac crest (L-3)	Perpendicular to the IR at the level of L-3	Zygapophyseal closest to the IR are best demonstrated
Lumbar Spine— PA Oblique Projection (RAO/LAO)	Patient semi-prone; rotate patient 45 degrees toward the affected side	Center the rotated spine to the ML of the grid device; adjust IR to a level 1.5 inches above iliac crest	Perpendicular to the IR at the level of L-3	Zygapophyseal farthest from the IR are best demonstrated
Sacrum—AP Projection	Patient supine	MSP centered to ML of grid device	15–25 degree cephalic angle entering a point 2 inches superior to the symphysis pubis	Sacrum free of superimposition
Coccyx—AP Projection	Patient supine	MSP centered to ML of grid device	10 degree caudal angle entering a point 2 inches superior to the symphysis pubis	Coccyx free of superimposition
Sacrum/Coccyx— Lateral Projection	Patient lateral recumbent on affected side	Center sacrum/coccyx to ML of the grid device	Perpendicular to the level of ASIS at a point 3.5 inches posterior; for individual coccyx, perpendicular and directed 3.5 inches posterior to ASIS and 2 inches inferior	Sacrum and coccyx in a lateral projection

Bending Left and Right (Weight-Bearing Method)

Bending films are performed in an erect position to evaluate the lumbar intervertebral disc spaces in their entirety. The central ray is directed to L-3 with a 15–20 degree caudal angle.

Scoliosis Series

Scoliosis surveys are performed in order to measure any abnormal lateral curvature of the vertebral column. They are imaged PA upright in order to reduce direct dosage to the reproductive organs. A 36-inch IR is utilized in order to image the spine in its entirety.

Pelvis

Pelvic Girdle

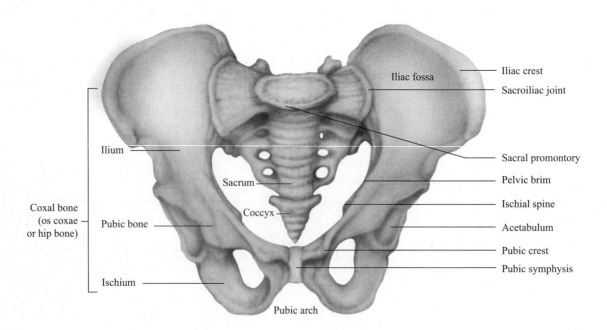

The bony pelvis consists of two innominate bones located on the lateral sides of the sacrum. Each individual innominate bone consists of three fused bones. They each contribute to the acetabulum, which is the area of articulation with the femoral head (hip joint). The ligament is found in the acetabulum, which connects the pelvis to the femoral head.

The bones of the pelvis are as follows:

Ilium: Located superiorly and contributes to two-fifths of the acetabulum. The medial auricular surface of these bones forms the sacroiliac joints bilaterally. The large expanded portion is referred to as the ala, or wing, the upper part of which is called the iliac crest. The iliac crest runs posteriorly from the anterior superior iliac spine (ASIS) to the posterior superior iliac spine (PSIS).

Ischium: Located posteriorly and inferiorly and contributes to two-fifths of the acetabulum. The ischial tuberosity is the thick, roughened inferior portion of the ischium which provides support when sitting.

Pubis: Located anteriorly and contributes to one-fifth of the acetabulum. It extends as a superior ramus anteriorly and medially. These rami from the right and left innominate bones meet at midline to constitute the symphysis pubis.

	Position of Patient	Position of Part	Central Ray	Structures
Pelvis—AP Projection	Patient in the supine position; MSP to ML of grid device	Medially rotate the feet 15–20 degrees; center the IR 2 inches inferior to ASIS and 2 inches superior to the symphysis pubis	Perpendicular to the IR	AP projection of pelvis and proximal one-third of the femur
Pelvis—AP Oblique Projection (Modified Cleaves Method)	Patient in the supine position; MSP to ML of grid device	Flex knees and hips and draw feet up as much as possible; abduct the thighs as much as possible; center IR 1 inch superior to the symphysis pubis	Perpendicular to enter the MSP 1 inch superior to the symphysis pubis	AP oblique projection of the femoral heads, necks, and trochanters
Anterior Pelvic Bones—AP Axial Outlet Projection	Patient in the supine position; MSP to ML of grid device	Center IR to coincide with the central ray	For males, 20–25 degree cephalic angle to a point 2 inches distal to the upper symphysis pubis; for females, 30–45 degree cephalic angle to a point 2 inches distal to the upper symphysis pubis	Pelvic rami
Acetabulum—AP Oblique Projection (Judet Method)	Patient in semi-supine position with affected side up; can also be performed with the affected side down	IR centered to affected hip; rotate body until the affected hip is elevated 45 degrees	Perpendicular to IR entering 2 inches inferior to the affected ASIS	Acetabular rim
SI Joints—AP Axial Projection	Supine	MSP to ML of grid device with no rotation of the pelvis; adjust IR to the CR	30–35 degree cephalic angle entering about 1.5 inches superior to the symphysis pubis on the MSP	Lumbosacral junction and SI joints free of superimposition

	Position of Patient	Position of Part	Central Ray	Structures
SI Joints—AP Oblique Projection (RPO/LPO)	Place patient in semi-supine position with affected side up	Elevate affected side 25–30 degrees; center IR to the level of ASIS; align body so that a sagittal plane 1 inch medial to elevated ASIS is aligned with the ML of the grid device	Perpendicular to the IR entering 1 inch medial to the elevated ASIS	SI joint farthest from the IR
SI Joints—PA Oblique Projection (RAO/LAO)	Place patient in semi-supine position with affected side down	Rotate affected side 25–30 degrees toward the table; center IR to the level of the ASIS; align body so that a sagittal plane 1 inch medial to ASIS closest is aligned with the ML of the grid device	Perpendicular to the IR entering 1 inch medial to the ASIS closest	SI joint closest to the IR
Hip—AP Projection	Patient in supine position with no rotation of the pelvis	Rotate the lower limb and foot 15–20 degrees internally unless contraindicated; center IR to femoral neck	Perpendicular to a point 2.5 inches distal to a line drawn perpendicular to the midpoint of a line between ASIS and symphysis pubis	Femoral head, neck, and the trochanters; proximal one-third or the femur
Hip—Axiolateral Projection (Danelius-Miller Method)	Patient in supine position	Flex knee of unaffected side and elevate thigh into a vertical position; support unaffected leg; place IR in vertical holder with upper border at iliac crest; angle lower border of IR until parallel with the femoral neck	Perpendicular to the long axis of the femoral neck entering mid-thigh and passing through the femoral neck	Acetabulum, head, neck, and trochanter of femur; considered the true lateral

	Position of Patient	Position of Part	Central Ray	Structures
Hip—AP Oblique Projection (Modified Cleaves Method)	Contraindicated in patients with suspected fracture; patient in the supine position; center the unaffected ASIS to the ML of the grid device	Flex affected knee and hip and abduct the thigh 45 degrees; center the IR to the femoral neck	Perpendicular to the femoral neck	AP oblique projection of the femoral head, neck, and trochanter
Hip—Axiolateral Projection (Clements-Nakayama Method)	Done when there are suspected bilateral hip fractures; patient supine with affected side on the edge of the table	IR in vertical holder with its lower margin below the patient; adjust so that it is parallel to the femoral neck and tilt back 15 degrees	15 degrees posterior angle and aligned perpendicular to the femoral neck	Lateral hip, acetabulum, femoral head, neck, and trochanter

Lower Extremity

The lower extremity consists of the foot, the lower leg, femur, and their corresponding joints.

Foot

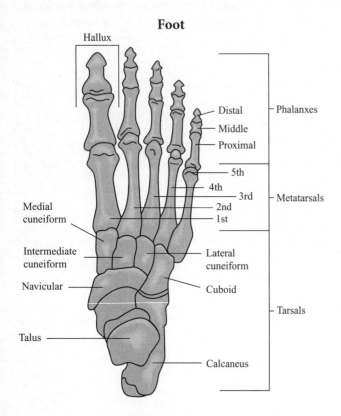

The foot contains 26 bones classified into groups as follows:

Fourteen phalanges: Each phalanx is classified as a long bone. The first toe, or hallux, contains two phalanges, the second through fifth toes all contain three phalanges.

Five metatarsals: Each metatarsal is classified as a long bone. They articulate with the phalanges distally and with the tarsals proximally. They make up the sole of the foot. The base of the fifth metatarsal projects laterally and is a common site of fracture.

Seven tarsals: The proximal portion of the foot. The distal row consists of the medial, intermediate, and lateral cuneiforms and is located between the metatarsals and the navicular. The cuboid is located on the lateral aspect and articulates distally with the fourth and fifth metatarsals and medially with both the navicular and the lateral cuneiform. The navicular (scaphoid) is located on the medial side of the foot and articulates distally to the cuneiforms and proximally with the talus. The talus (astragulus) is the tarsal bone which articulates with the tibia superiorly to form the ankle joint. Inferiorly, it articulates with the calcaneous (os calcis), of which its tuberosity forms the heel, to form the three-faced subtalar joint. A canal termed the *sinus tarsi* is present on this joint.

Ankle joint

The ankle joint, or mortise, is formed from the articulations between the superior aspect of the talus and the medial malleolus, and the inferior surface of the tibia and the lateral malleolus of the fibula.

Lower Leg

Tibia/Fibula

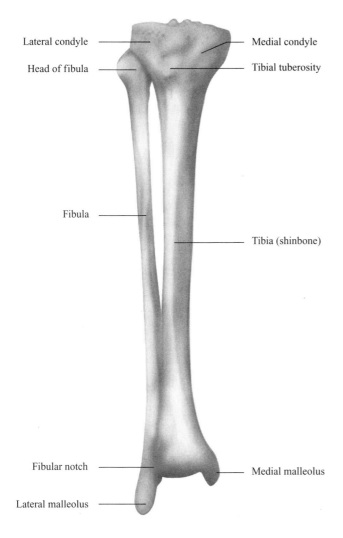

Lateral condyle

Head of fibula

Fibula

Fibular notch

Lateral malleolus

Medial condyle

Tibial tuberosity

Tibia (shinbone)

Medial malleolus

The lower leg is made up of the tibia (medial bone) and fibula (lateral bone). The tibia consists of a shaft and two extremities and articulates with the femur superiorly, the talus inferiorly, and the fibula laterally. The distal articular surface of the tibia is smooth in order to articulate with the talus below. The medial aspect of the distal tibia is termed the *medial malleolus* and is palpable, the lateral aspect contains a notch to articulate with the fibula (distal tibiofibular

joint). The proximal aspect of the bone presents two eminences on each side termed the *medial* and *lateral condyle*, which articulate with the condyles of the femur. Between these two surfaces is the intercondyloid eminence (tibial spine). The lateral aspect contains a roughened surface to articulate with the fibula (proximal tibiofibular joint). On the anterior aspect of the proximal tibia, an elevation in the bone is present called the tibial tuberosity, which is a site of ligament insertion.

The fibula is lateral to the tibia and consists of a shaft with two extremities. The proximal portion is termed the *head* and articulates with the lateral condyle of the tibia. Just distal to the head is a slender portion called the neck. The distal end of the fibula is the lateral malleolus, which articulates with the tibia as well as helps form the ankle joint.

Knee

The knee joint is formed by the articulation between the condyles of the proximal tibia, the condyles of the distal femur, and the patella.

Femur

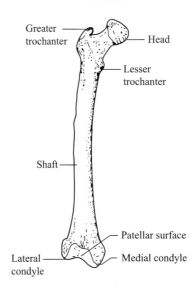

Greater trochanter

Head

Lesser trochanter

Shaft

Patellar surface

Lateral condyle

Medial condyle

The femur is the largest bone in the body. The distal end of the femur, which articulates with the tibia, consists of two large condyles (medial and lateral) separated by a posterior depression called the intercondyloid fossa. Immediately superior to the condyles are the smaller epicondyles. On the anterior surface, there is an area for articulation with the patella. On the posterior shaft of the femur, a ridge exists called the linea aspera. On the proximal end, the rounded head of the femur articulates with the acetabulum of the pelvis in a typical ball and socket presentation. The fovea capitis on the head is for ligament attachment. Below the head is the femoral neck, which ascends at a 120 degree angle from the shaft. Just distal to the neck are large processes on the posterior aspect called the greater and lesser trochanters. The intertrochanteric line runs on the anterior surface between both trochanters. The intertrochanteric crest runs posteriorly. The greater trochanter is palpable and is a common landmark for positioning.

Hip

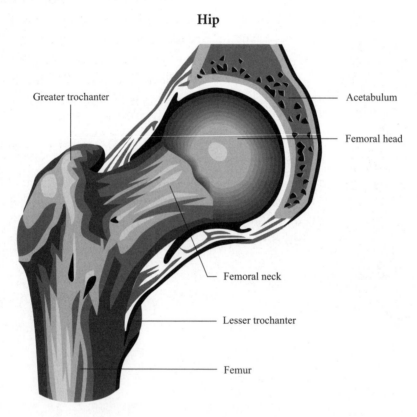

Anterior Femur

The hip joint is the articulation between the femoral head and the acetabulum of the pelvis. What are known as common hip fractures normally take place on the neck of the femur.

Positions

	Position of Patient	Position of Part	Central Ray	Structures
Toes— AP Projection	Seated or supine with affected knee flexed and sole of foot resting on table	Center toes over IR	Perpendicular through the third metatarsophalangeal joint; for open joint spaces 15 degrees posteriorly through the third metatarsophalangeal joint	14 phalanges of the toes, distal metatarsals; when angled metatarsophalangeal joints
Toes— AP Oblique Projection Medial Rotation	Seated or supine with affected knee flexed and sole of foot resting on table	Center toes over IR; medially rotate the foot and leg to form a 30–45 degree angle to the IR	Perpendicular through the third metatarsophalangeal joint	14 phalanges of the toes; distal metatarsals rotated medially
Toes— Lateral Projection (Mediolateral/ Lateromedial)	Lateral recumbent; tape the toes above the ones being examined	Place the toe into a true lateral position utilizing the mediolateral or lateromedial, depending on the toe and its relation to the IR to reduce the magnification	Perpendicular through the metatarsophalangeal joint for the great toe; the proximal interphalangeal joints for the other four toes	Lateral projection of the toe(s); interest as free of superimposition as possible
Sesamoids— Tangential Projection (Lewis Method)	Patient in prone position and knee bent slightly	Rest the hallux on the table in a position of dorsiflexion, place ball of foot perpendicular to the horizontal; center IR to the second metatarsal	Perpendicular and tangential to the first metatarsophalangeal joint	Sesamoids free of superimposition
Sesamoids— Tangential Projection (Causton Method)	Lateral recumbent on the unaffected side; flex knee	Place the foot into a lateral position; the first metatarsophalangeal joint should be perpendicular; place the IR under the distal metatarsal region	40 degree angle toward the heel directed to the ball of the foot	Axiolateral projection of the sesamoids

	Position of Patient	Position of Part	Central Ray	Structures
Foot—AP Axial Projection	Seated or supine with affected knee flexed and sole of foot resting flat on table	IR under affected foot centered to the base of the third metatarsal	10 degree angle toward the heel, toward the base of the third metatarsal	14 phalanges, navicular, cuneiforms, cuboid; talus and calcaneous are not well visualized
Foot— AP Oblique Projection Medial Rotation	Seated or supine with affected knee flexed and sole of foot resting flat on table	Center foot over IR; medially rotate the foot and leg to form a 30 degree angle to the IR	Perpendicular to the base of the third metatarsal	Sinus tarsi, tuberosity of the base of the fifth metatarsal, interspaces involving the lateral side of the foot
Foot— AP Oblique Projection Lateral Rotation	Seated or supine with affected knee flexed and sole of foot resting flat on table	Center foot over IR; laterally rotate the foot and leg to form a 30 degree angle to the IR	Perpendicular to the base of the third metatarsal	Interspaces on the medial side of the foot
Foot—Lateral Projection (Mediolateral/ Lateromedial)	Mediolateral—lying on table, turn towards affected side until foot and leg are lateral; lateromedial—lying on table, turn toward unaffected side until foot and leg are lateral	Center IR to midarea of the foot; dorsiflex foot so that plantar surface is perpendicular to the IR	Perpendicular to the base of the third metatarsal	Superimposed phalanges and metatarsals, ankle joint superimposed, tarsals superimposed; it should be noted that the lateromedial, while more difficult for the patient, is the truer lateral of the foot
Foot—Lateral Projection (Lateromedial Weight-Bearing Method)	Patient upright	Place IR in slot between patient's feet; patient must equally bear weight; IR should be centered to the midarea of the foot	Perpendicular to just above the base of the third metatarsal	Lateromedial image of the foot while weight-bearing to best demonstrate the longitudinal arch

	Position of Patient	Position of Part	Central Ray	Structures
Foot—AP Projection (Axial Weight- Bearing Method)	Standing upright	IR under affected foot centered to the base of the third metatarsal	10 degree angle toward the heel and the base of the third metatarsal, though more angle may be needed; for composite view, a controlled double exposure where the first projection is made with a 15 degree angle toward the heel and the second projection is made while not moving the foot using the 25 degree angle to the posterior surface of the ankle	14 phalanges, navicular, cuneiforms, cuboid; talus and calcaneous are not well visualized; composite view shows all of the bones of the foot without superimposition
Calcaneous— Lateral Projection (Mediolateral)	Lying on table, turn toward affected side until foot and leg are lateral	IR centered to the calcaneous	Perpendicular to the calcaneous 1 inch distal to the medial malleolus	Lateral profile of calcaneous and ankle
Calcaneous— Axial Projection (Dorsoplantar)	Prone	Elevate leg on support; dorsiflex ankle until the foot is perpendicular to the table; IR against plantar surface	40 degree caudal angle entering the dorsal surface of the ankle	Calcaneous and subtalar joint
Calcaneous— Axial Projection (Plantodorsal)	Supine with leg extended	IR under ankle; place a radiolucent strap around ball of foot and have patient pull until foot is at right angle dorsiflexion	40 degree cephalic angle entering the base of the third metatarsal	Calcaneous and subtalar joint
Ankle—AP Projection	Supine, leg fully extended	Center IR to ankle joint; flex ankle so that foot is vertical	Perpendicular through the ankle joint midway between the malleoli	Ankle joint, distal tibia/fibula, proximal talus

	Position of Patient	Position of Part	Central Ray	Structures
Ankle— Lateral Projection (Mediolateral/ Lateromedial)	Supine, turn leg toward the affected side until ankle is lateral for mediolateral; turn leg toward the unaffected side until affected ankle is lateral for lateromedial	Center IR to lateral surface of ankle joint; dorsiflex foot	Perpendicular entering the malleolus	Lateral ankle joint, tibia/fibula, talus, calcaneous
Ankle—AP Oblique Projection (Medial Rotation)	Supine, leg fully extended	Center IR to ankle joint; flex ankle so that foot is vertical; rotate the leg and foot medially 45 degrees	Perpendicular through the ankle joint midway between the malleoli	Distal ends of tibia and fibula; distal tibiofibular joint best demonstrated
Ankle Mortise— AP Oblique Projection (Medial Rotation)	Supine, leg fully extended	Center IR to ankle joint; flex ankle so that foot is vertical; rotate the leg and foot medially 15–20 degrees	Perpendicular through the ankle joint midway between the malleoli	Entire ankle mortise joint open and free of any superimposition
Tibia/Fibula (Leg)–AP Projection	Supine, leg fully extended	IR centered to include both ankle and knee joints if possible; ensure foot fully vertical and femoral condyles are parallel to the IR	Perpendicular to center of leg	Tibia, fibula, knee, and ankle joints
Tibia/Fibula (Leg)—Lateral Projection	Supine	IR centered to include both ankle and knee joints if possible; turn body toward the affected side until patella is perpendicular to IR	Perpendicular to center of leg	Tibia, fibula, knee, and ankle joints

	Position of Patient	Position of Part	Central Ray	Structures
Knee—AP Projection	Supine; ensure femoral condyles are parallel to the plane of the table	Center IR one-half inch distal to the patellar apex	Directed to a point one-half inch distal to patellar apex; if ASIS measures less than 19 cm, 3–5 degree caudal angle; if between 19 and 24 cm, no angulation is required; if greater than 24 cm, 3–5 degree cephalic angle	Knee joint, superimposed patella overlying femur
Knee—PA Projection	Prone; foot dorsiflexed so toes rest on table; ensure no rotation of the femoral condyles	IR centered to a point one-half inch below patellar apex	5 degree caudal angle to exit one-half inch below patellar apex	Knee joint, less magnified superimposed patella overlying femur
Knee—Lateral Projection	Supine; turn patient onto affected side, place the unaffected knee behind	Flex knee 20–30 degrees, adjust epicondyles of the femur until they are perpendicular to the IR; patella will be perpendicular to the IR	5–7 degree cephalic angle directed 1 inch distal to the medial epicondyle	Knee joint free of superimposition of the medial condyle due to CR angulation, distal femur, proximal tibia, open patellofemoral joint space
Knee—AP Projection (Weight-Bearing)	Erect; ensure femoral condyles are parallel to the plane of the grid device	Center IR one-half inch distal to the patellar apex	Horizontal and perpendicular to the IR one-half inch below apex of patella	Best evaluation of the knee joint space
Knee—AP Oblique Projection (Lateral Rotation)	Supine	Center IR one-half inch distal to the patellar apex; externally rotate femur and leg 45 degrees	Directed to a point one-half inch distal to patellar apex; if ASIS measures less than 19 cm, 3–5 degree caudal angle; if between 19 and 24 cm, no angulation is required; if greater than 24 cm, 3–5 degree cephalic angle	Laterally rotated femoral condyles; lateral margin of the patella free of superimposition; fibula superimposed by the tibia

	Position of Patient	Position of Part	Central Ray	Structures
Knee—AP Oblique Projection (Medial Rotation)	Supine	Center IR one-half inch distal to the patellar apex; elevate affected hip slightly; internally rotate femur and leg 45 degrees	Directed to a point one-half inch distal to patellar apex; if ASIS measures less than 19 cm, 3–5 degree caudal angle; if angle is between 19 and 24 cm, then no angulation is required; if greater than 24 cm, then 3–5 degree cephalic angle	Best demonstration of the proximal tibia/fibula joint and fibula head; medial margin of patella free of superimposition
Knee—PA Axial Projection (Holmblad Method)	Standing with affected knee flexed and resting on raised support	IR against anterior surface of the knee and centered to patella; flex knee 70 degrees from full extension	Perpendicular to the lower leg entering midpoint of IR	Intercondylar fossa and tubercles of the tibial eminences
Knee—PA Axial Projection (Camp-Coventry Method)	Prone, no rotation of leg	Flex affected knee to a 40 or 50 degree angle; center upper border of IR to knee joint	Perpendicular to lower leg (angled 40 degrees when knee is flexed 40 degrees or 50 degrees when knee is flexed 50 degrees)	Unobstructed intercondyloid fossa and tubercles of the tibial eminences
Patella—Lateral Projection	Lateral recumbent	Turn on affected side and flex affected knee 5–10 degrees; ensure condyles are perpendicular to IR	Perpendicular entering midpatellofemoral joint with tight collimation	Lateral patella and patellofemoral joint space
Patella—Tangential Projection (Merchant Method)	Supine with both knees toward end of table; support knees and legs using holding device	Femora parellel with tabletop; adjust flexion of device so knees form a 40 degree angle; ensure no rotation, rest IR on patient's shins 1 foot distal to patellae	30 degree angle caudal from the horizontal entering midway between patellae at the level of the patellofemoral joint	Nondistorted patellae and patellofemoral joints

	Position of Patient	Position of Part	Central Ray	Structures
Patella—Tangential Projection (Settegast Method)	Contraindicated if transverse fracture of the patella has not been ruled out; supine	Flex knee as much as possible; have patient hold IR in place with the upper margin above the flexed knee	Perpendicular to the joint space when the joint is perpendicular; collimate tightly	Patella, vertical fractures, patellofemoral articulation
Patella—Tangential Projection (Hughston Method)	Prone; ensure no rotation of knee	IR directly under patient's affected knee; flex so that lower leg forms a 50–60 degree angle from the table surface; rest the foot on the collimator box	45 degree angle cephalic, directed toward the patellofemoral joint	Patella, good position for subluxation of patella
Femur—AP Projection	Supine; may be necessary to image in two parts to include both hip and knee joints	Affected femur to ML of the IR; for distal femur, place epicondyles parellel with the IR, lower margin of the IR 2 inches below the knee joint; for proximal femur, place upper margin of IR at the level of ASIS; internally rotate the leg 10–15 degrees	Perpendicular to the midfemur	AP projection of femur including both hip and knee joints
Femur—Lateral Projection (Mediolateral)	Turn patient onto affected side; center affected femur to midline of table	For distal femur, with the pelvis in a true lateral position, roll the unaffected leg forward and flex the affected knee 45 degrees; lower margin of IR 2 inches distal to knee joint For proximal femur, upper margin of IR at the level of ASIS, roll the pelvis onto the affected side approximately 15 degrees, and brace the unaffected leg out of the field of view	Perpendicular to the midfemur	Lateral projection of the femur

Upper Extremity
Hand

Hand and Upper Extremity

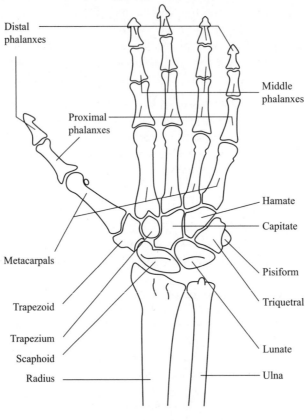

Distal phalanxes

Middle phalanxes

Proximal phalanxes

Metacarpals

Hamate

Capitate

Pisiform

Triquetral

Trapezoid

Trapezium

Scaphoid

Lunate

Radius

Ulna

Bones of the Hand

There are 27 bones in the normal hand, classified as follows:

Fourteen phalanges: the fingers. The first finger, or thumb, consists of two phalanges: a proximal and a distal. The second and third fingers consist of three phalanges: a proximal, a middle, and a distal. The articulations between phalanges are termed interphalangeal joints.

Five metacarpals: the palm of the hand. The head of each metacarpal (knuckle) articulates with the base of the corresponding phalanx. This articulation is called the metacarpophalangeal joint.

Eight carpal bones: the wrist. The bones are arranged in two rows: proximal and distal. The proximal row from medial to lateral is as follows:

- Navicular (scaphoid)—most frequently fractured carpal
- Lunate (semilunar)
- Triquetrum (triangular)
- Pisiform—partially superimposed by the triquetrum

The distal row from medial to lateral is as follows:

- Trapezium (greater multangular)
- Trapezoid (lesser multangular)
- Capitate (os magnum)—largest carpal
- Hamate (unciform)—contains characteristic hooklike process

The wrist joint is the articulation between the proximal row of carpals and the distal ends of the radius (majority) and ulna. What are known as common wrist fractures sometimes refer to fractures of the distal radius.

Radius and Ulna

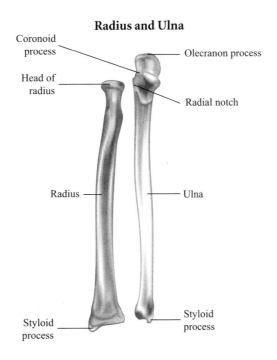

Radius and Ulna

The radius (medially) and ulna (laterally) compose the forearm. The ulna is the larger of the two bones and is wider on the proximal end and more slender on its distal end. The radius is slender on its proximal end and broader on its distal end. The distal end of the ulna consists of the head and styloid process. On the proximal end of the ulna, two processes, the olecranon process (located on the posterior aspect) and the coronoid process (anterior aspect), are separated by the trochlear (semilunar) notch. It is here that the ulna articulates with the humerus. Lateral and distal to the trochlear notch is the area for articulation with the proximal radius called the radial notch.

The distal radius presents distally a styloid process on its lateral border. Medially, the ulna notch provides an articulation for the ulnar head. The proximal head of the radius is cylindrical and fits into the radial notch of the ulna. Inferior to the head is the neck, followed by an elevation termed the radial tuberosity.

Elbow

Elbow (Anterior and Posterior View)

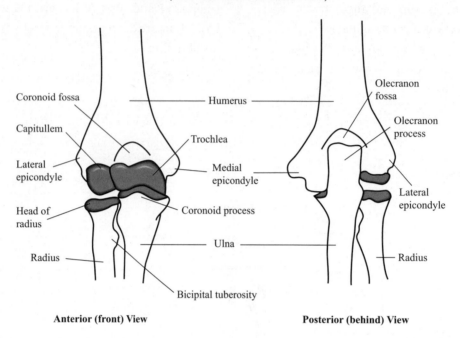

Coronoid fossa

Capitullem

Lateral epicondyle

Head of radius

Radius

Humerus

Trochlea

Medial epicondyle

Coronoid process

Ulna

Bicipital tuberosity

Olecranon fossa

Olecranon process

Lateral epicondyle

Radius

Anterior (front) View

Posterior (behind) View

The elbow is the joint formed by the articulation between the proximal radius and ulna with the distal end of the humerus.

Humerus

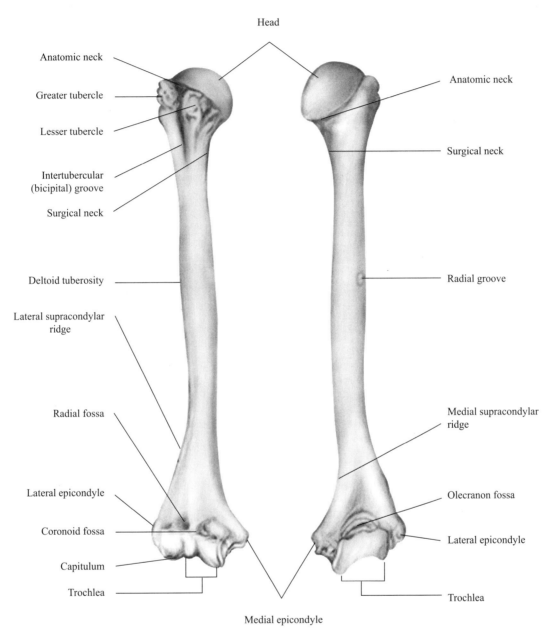

Humerus

Head

Anatomic neck

Greater tubercle

Lesser tubercle

Intertubercular (bicipital) groove

Surgical neck

Deltoid tuberosity

Lateral supracondylar ridge

Radial fossa

Lateral epicondyle

Coronoid fossa

Capitulum

Trochlea

Medial epicondyle

Anatomic neck

Surgical neck

Radial groove

Medial supracondylar ridge

Olecranon fossa

Lateral epicondyle

Trochlea

The lateral aspect of the distal humerus is called the capitulum, which articulates with the superior part of the radial head. The medial aspect is called the trochlea. This articulates with the trochlear (semilunar) notch. Lying just above the capitulum and trochlea are medial and lateral epicondyles. The posterior distal humerus also contains a small depression that receives the olecranon process of the ulna when the arm is in extension. This is called the olecranon fossa. The proximal humerus contains a large, rounded head, which articulates with the scapula at the glenoid fossa. The anatomical neck is the area where the head and the body of the humerus meet. The greater tubercle is located on the anterior lateral aspect, while the lesser tubercle is located on the anterior medial aspect of the proximal humerus. Located between the two tubercles is the bicipital groove. The slimmer area of the shaft just inferior to the two tubercles is called the surgical neck. This is the location of common humeral fractures.

Shoulder Girdle

Shoulder

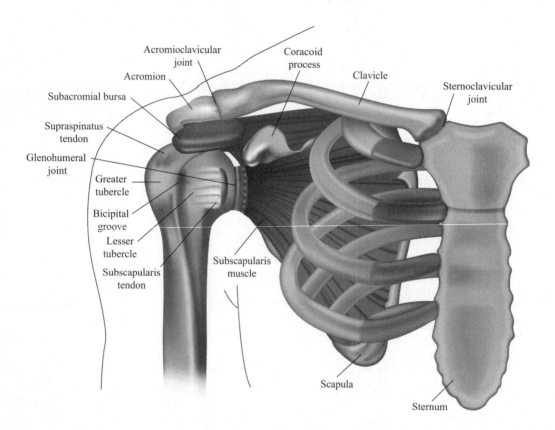

Shoulder Joint

The shoulder joint is made up of the articulation between the humeral head and the glenoid cavity of the scapula. There are various muscles present around this area, which are termed the rotator cuff.

Scapula

The scapula is a triangular, flat bone. It has three borders: superior border, vertebral (or medial) border, and axillary (or lateral) border. Where the vertebral and axillary borders meet inferiorly is termed the inferior angle, which is a palpable landmark and lies at the level of T-7. The most lateral aspect of the axillary border is the glenoid fossa, which articulates with the humeral head. The acromion process is part of the scapular spine, which projects laterally and is located on the posterior surface of the scapula and divides the scapula into infraspinatus and supraspinatus fossa. It is here that some muscles of the shoulder articulate. On the superior border anteriorly, the palpable coracoid process is present, which is useful in positioning of the shoulder.

Clavicle

The clavicle is located on the anterior upper surface of the thorax. It is S-shaped and lies on an oblique plane to the thorax. It has two articulating ends: medially with the sternum and laterally with the acromion of the scapula. It is frequently fractured early in life, especially during birth.

AC Joints

The acromioclavicular joint is the articulation between the acromion of the scapula and the clavicle. It is a frequent site of dislocation, especially in weight lifters.

Positions

	Position of Patient	Position of Part	Central Ray	Structures
Fingers (Second through Fifth)— PA Projection	Patient seated, elbow flexed 90 degrees, forearm and hand on same transverse plane	Palm side down place separated digits on IR; center the affected proximal interphalangeal joint (PIP) to the IR	Perpendicular to the PIP of the affected digit	Entire phalange with its joints and head of the metacarpal
Fingers (Second through Fifth)— Lateral Projection	Patient seated, elbow flexed 90 degrees, forearm and hand on same transverse plane	Close all unaffected digits into a fist, and place the affected digit as close to the IR as possible; adjust IR to ML of affected PIP joint	Perpendicular to the PIP of the affected digit	Entire phalanx with its joints and head of the metacarpal in a lateral position
Fingers (Second through Fifth)— PA Oblique Projection	Patient seated, elbow flexed 90 degrees, forearm and hand on same transverse plane	Adjust IR to affected PIP joint; externally rotate the hand until it forms a 45 degree angle with the IR	Perpendicular to the PIP of the affected digit	Entire phalanx with its joints and head of the metacarpal in an oblique position
Thumb (first digit)— AP Projection	Patient seated, elbow flexed 90 degrees, forearm and hand on same transverse plane	Place hand in extreme internal rotation until the thumb attains a true AP position on the IR; center IR to MCP joint	Perpendicular to the metacarpophalangeal joint (MCP)	AP projection of the 2 phalanges of the thumb, open joint spaces, first metacarpal head
Thumb—PA Oblique Projection	Patient seated, elbow flexed 90 degrees, forearm and hand on same transverse plane	Place hand palm side down on the surface of the IR; center IR to MCP joint	Perpendicular to the metacarpophalangeal joint	Oblique projection of the two phalanges of the thumb, open joint spaces, first metacarpal head

	Position of Patient	Position of Part	Central Ray	Structures
Thumb—Lateral Projection	Patient seated, elbow flexed 90 degrees, forearm and hand on same transverse plane	Place hand palm side down on the surface of the IR; center IR to the MCP joint; adjust hand until the thumb is in a true lateral position	Perpendicular to the metacarpophalangeal joint	Lateral projection of the two phalanges of the thumb, open joint spaces, first metacarpal head
Hand—PA Projection	Patient seated, elbow flexed 90 degrees, forearm and hand on same transverse plane	Place hand palm side down on the IR; center IR to the MCP joints; spread fingers	Perpendicular to the third MCP joint	PA projection of digits two through five, five metacarpals, eight carpals, all related joint spaces, oblique projection of thumb
Hand—PA Oblique Projection	Patient seated, elbow flexed 90 degrees, forearm and hand on same transverse plane	Place hand palm side down on the IR; rotate hand laterally until it forms a 45 degree angle with the plane of the IR	Perpendicular to the third MCP joint	Oblique projection of all bones of the hand with slight overlap by the medial metacarpals
Hand—Lateral Projection	Patient seated, elbow flexed 90 degrees, forearm and hand on same transverse plane with the hand in a lateral position with the ulnar side down	Extend digits, center the IR to the MCP joints; if necessary, fan the fingers out to avoid superimposition of the digits	Perpendicular to the second MCP joint	Extended lateral— soft tissue for foreign bodies; fan removes superimposition of the digits
Wrist—PA Projection	Patient seated, elbow flexed 90 degrees, forearm and hand on same transverse plane	Hand palmar side down, center wrist to the IR; flex the digits to bring the carpals closer to the IR	Perpendicular to the midcarpal area	Eight carpal bones, proximal metacarpals, distal radius, and ulna Note: AP view is best for carpal interspaces
Wrist—Lateral Projection	Patient seated, elbow flexed 90 degrees, forearm and hand on same transverse plane with the hand in a lateral position with the ulnar side down	Center IR to carpal region; ensure that hand is in a true lateral position	Perpendicular to the wrist joint	Lateral projection of the carpals; best used for anterior or posterior displacements

	Position of Patient	Position of Part	Central Ray	Structures
Wrist—PA Oblique Projection (Lateral Rotation)	Patient seated, elbow flexed 90 degrees, forearm and hand on same transverse plane	Hand palmar side down, center wrist to the IR; pronate wrist until it forms a 45 degree angle with the IR	Perpendicular to the midcarpal area	Trapezium and scaphoid
Wrist—AP Oblique Projection (Medial Rotation)	Patient seated, elbow flexed 90 degrees, forearm and hand on same transverse plane	Hand dorsal side down; center wrist to the IR; supinate wrist until it forms a 45 degree angle with the IR	Perpendicular to the midcarpal area	Pisiform free of superimposition, medial carpal bones
Wrist—PA Projection (Ulnar Deviation)	Patient seated, elbow flexed 90 degrees, forearm and hand on same transverse plane	Hand palmar side down; center wrist to the IR; turn hand outward while stabilizing the wrist to bring the joint into ulnar deviation	Perpendicular to the scaphoid	Scaphoid without foreshortening
Wrist—PA Axial Projection, Stecher Projection	Patient seated, elbow flexed 90 degrees, forearm and hand on same transverse plane	Elevate the IR 20 degrees off the table; place the wrist in the PA position on the IR; if the IR cannot be elevated, the CR can be angled 20 degrees toward the elbow	Perpendicular to the scaphoid	Scaphoid with open articulations
Carpal Canal–Tangential Projection (Gaynor-Hart Method)	Seated, extend forearm along table	With the palm down, hyperextend the wrist to as close to vertical as possible; center IR to the level of the radial styloid process	25–30 degree angle to the long axis of the hand directed 1 inch distal to the base of the third metacarpal	Carpal canal, hook of the hamulus, pisiform
Forearm—AP Projection	Seated, entire arm extended on same transverse plane	Supinate hand until true AP position of forearm is achieved; adjust IR to long axis of forearm	Perpendicular to the midpoint of the forearm	Radius and ulna, both wrist and elbow joints

	Position of Patient	Position of Part	Central Ray	Structures
Forearm—Lateral Projection	Patient seated, elbow flexed 90 degrees, forearm and hand on same transverse plane	Center IR to long axis of forearm; adjust forearm into a true lateral position	Perpendicular to the midpoint of the forearm	Radius and ulna (superimposed distally), superimposed epicondyles of humerus, lateral wrist and elbow
Elbow—AP Projection	Seated, entire arm extended on same transverse plane	Center IR to elbow joint; extend arm fully and rotate the hand until the palm is facing up; rotate the patient laterally until the humeral epicondyles are parallel to the IR	Perpendicular to the elbow joint	Elbow joint, radial head, neck, and tuberosity slightly superimposed by the ulna
Elbow—Lateral Projection	Patient seated, elbow flexed 90 degrees, forearm and hand on same transverse plane	Center IR to flexed elbow joint	Perpendicular to the elbow joint	Olecranon process in profile, elbow joint, distal humerus, and proximal forearm in lateral position
Elbow—AP Oblique Projection (Medial Rotation)	Seated, entire arm extended on same transverse plane	Center IR to elbow joint, extend arm fully; medially rotate the hand until the palm is flat on the table. Adjust elbow to place the anterior surface at a 45 degree angle	Perpendicular to the elbow joint	Coronoid process in profile, olecranon lying in the olecranon fossa, oblique elbow
Elbow—AP Oblique Projection (Lateral Rotation)	Seated, entire arm extended on same transverse plane	Center IR to elbow joint, extend arm fully; with palm side up, externally rotate the hand to place the elbow at a 45 degree angle	Perpendicular to the elbow joint	Radial head, neck, and tuberosity free of superimposition from the ulna

	Position of Patient	Position of Part	Central Ray	Structures
Elbow—AP Projection Partial Flexion (Distal Humerus and Proximal Forearm)	Seated; for distal humerus, entire humerus on same plane; for proximal forearm, entire forearm with hand supinated on same plane	For distal humerus, with forearm both in flexion and elevated off the table, center IR to the elbow joint; for proximal forearm, place entire forearm flat on table with elbow in flexion and hand supinated	Perpendicular to the elbow joint for both distal humerus and proximal elbow	Two views are necessary if the elbow cannot be flexed due to patient
Humerus—AP Projection	Upright	Adjust upper margin of IR so that it lies 1.5 inches above the humeral head; minimally abduct arm and supinate hand, ensuring epicondyles of humerus are parallel with the IR	Perpendicular to the midpoint of humerus	Greater tubercle of the humeral head in profile laterally; entire humerus including shoulder and elbow joints
Humerus—Lateral Projection	Upright	Adjust upper margin of IR so that it lies 1.5 inches above the humeral head; have patient flex the elbow 90 degrees and internally rotate the arm until epicondyles are perpendicular to the IR	Perpendicular to the midpoint of humerus	Lesser tubercle in profile medially; entire humerus including shoulder and elbow joints
Shoulder—AP Projection (Internal and External Rotations)	Upright or supine	Adjust center of IR so that it lies 1 inch inferior to the coracoid process; for internal rotation, minimally abduct arm and supinate hand, ensuring epicondyles of humerus are parallel to the plane of the IR; for external rotation, internally rotate the arm until epicondyles are perpendicular to the IR	Perpendicular to a point 1 inch inferior to the coracoid process	Internal, greater tubercle of the humeral head in profile laterally; external, lesser tubercle in profile medially

	Position of Patient	Position of Part	Central Ray	Structures
Shoulder/Proximal Humerus—Transthoracic Lateral Projection (Lawrence Method)	Upright, patient in lateral position with affected side closest to grid device	Raise unaffected arm until it is out of the path of the central ray; MCP perpendicular to the IR; center IR to the surgical neck of the affected humerus	Perpendicular to the IR exiting the surgical neck of the affected side	Proximal humerus visualized through the lung field
Shoulder—Inferosuperior Axial Projection (Lawrence Method)	Supine with upper part of the body elevated via support off the table	Abduct affected arm until it achieves a 90 degree angle with the body; externally rotate the humerus; IR in vertical device against the superior surface of the affected shoulder with its lateral margin touching the patient's neck	Horizontal through the axilla 15–30 degrees medial angulation	Scapulohumeral joint relationship
Shoulder/Glenoid—AP Oblique Projection (Grashey Method)	Upright, affected arm flexed at elbow and resting on patient's waist	IR centered to shoulder joint; oblique patient 35–45 degrees toward the affected side	Perpendicular to the gleno-humeral joint; two inches inferior and two inches medial to the lateral border of the shoulder	Glenoid cavity in profile
Shoulder/Humerus—PA Oblique Projection, Scapular "Y" View	Upright, facing the grid device	Place affected shoulder toward grid device; rotate patient until MCP is 45–60 degrees to the IR; center IR to shoulder joint	Perpendicular to the shoulder joint	Humeral head superimposed over the scapular "Y"; position is for anterior and posterior dislocations of the humeral head
Shoulder—Tangential Projection (Neer Method)	Upright, facing the grid device	Place affected shoulder toward grid device; rotate patient until MCP is 45–60 degrees to the IR; center IR to shoulder joint	10–15 degrees caudal angle entering the humeral head superiorly	Shoulder outlet

	Position of Patient	Position of Part	Central Ray	Structures
Scapula—AP Projection	Upright or supine	Affected scapula centered to ML of grid device; abduct arm 90 degrees; adjust upper margin of IR two inches above surface of shoulder	Perpendicular to the midscapular area 2 inches below coracoids	Lateral aspect of scapula free of superimposition
Scapula—Lateral Projection (RAO/LAO)	Upright facing the grid device	Place affected shoulder toward grid device; rotate patient until MCP is 45–60 degrees to the IR to place the scapular body perpendicular to the IR	Perpendicular to the midmedial border of the affected scapula	Lateral scapula
Clavicle—AP Projection	Supine or upright, relax shoulders	IR centered to affected midclavicle	Perpendicular to the midshaft of clavicle	Clavicle lying on oblique plane
Clavicle—AP Lordotic	Upright standing 1 foot in front of vertical grid device or supine	Patient leans back in lordotic position; center IR to clavicle	If upright lordotic, 0–15 degrees cephalic angle directed to clavicle; if supine, 15–30 degrees cephalic angle to clavicle	Clavicle projected free of superimposition
Clavicle—PA Projection	Prone	Center IR to clavicle	15–30 degrees caudal angle directed to clavicle	Clavicle projected free of superimposition
AC Joints–AP Projection (Pearson Method)	Patient standing erect in front of grid device with arms at side	Two separate exposures; IR adjusted to height at level of AC joints; MSP of body to ML of grid First exposure: patient holding equal weight in both hands; second exposure: patient natural	SID 72 inches for both exposures, perpendicular to the ML of the body at level of AC joints	AC joints bearing weight and non-weight-bearing

Skull

The skull is made up of two parts: the cranium and the facial bones. The bones of the skull are separated by sutures. The coronal suture separates the frontal and parietal bones. The sagittal suture separates the two parietal bones. The lamboidal suture separates the occipital and parietal bones. The squamosal suture separates the parietal and temporal bones. Where the sagittal and coronal sutures near the vertex (or top) of the skull is termed the bregma. Where the sagittal and lamboidal sutures meet on the posterior skull is called the lambda.

Cranium: The cranium is made up of eight bones and provides protection for the brain.

Frontal bone: Commonly known as the forehead. Contains the glabella, a useful landmark for CT, which is a smooth prominence located between the eyebrows. There are two frontal sinuses associated with the frontal bone.

Occipital bone: The most posterior and inferior bone of the cranium. Inferiorly located is the foramen magnum, an opening where the brain stem extends downward.

Temporal bones (2): Located on both lateral sides of the cranium. Contains a squamous portion, tympanic portion, styloid process, zygomatic process, petrous portions, and mastoid process. The mastoid process contains air cells that assist, along with the petrous portions, in balance.

Ethmoid bone: Found between the orbits and articulates with the frontal and sphenoid bones. It consists of the cribiform plate horizontally, the perpendicular plate, which is vertical, and two air-filled spongy lateral masses called labyrinths on each side. The crista galli projects superiorly on the cribiform plate.

Parietal bones (2): Forms part of the lateral cranium. It articulates with the frontal, temporal, occipital, sphenoid, and opposite parietal bones.

Sphenoid bone: The sphenoid is found on the base of the cranium. It consists of a body, a pair of lesser and greater wings, and a pair of pterygoid processes inferiorly. The body contains the sphenoid sinuses. The sella turcica lies on the superior aspect and provides a housing for the pituitary gland.

Facial bones: There are 14 facial bones.

Nasal bones (2): Small and thin bones that form the bridge of the nose.

Lacrimal bones (2): The smallest bones in the skull, they are located between the ethmoid and the maxilla.

Maxillary bones (2): Largest facial bones. The maxillary bones contain the large maxillary sinus and the infraorbital foramen for the optic nerve passageway.

Zygomatic bones (2): These are the cheeks. They unite with the maxilla to form the zygomatic arch.

Palantine bones (2): These bones help form the roof of the mouth (bony palate) and the orbit.

Inferior nasal conchae (2): These extend horizontally along the lateral wall of the nasal cavity.

Vomer: This forms the inferior part of the nasal septum.

Mandible: The jaw, it is the largest bone of the face. It consists of a body, two angles, and two vertical rami, one on each side. On the rami superiorly there are two processes: the coronoid and the condylar. The condylar articulates with the temporal bone to form the TMJ.

Positions

It is important to note that several positions for the skull anatomy can be utilized for varying protocols without change in position or central ray. For example, the Waters position for facial bones can also be utilized for sinuses or orbits. Positioning protocols vary based on radiologist preferences.

	Position of Patient	Position of Part	Central Ray	Structures
Skull—Lateral Projection	Patient upright or semi-prone; in trauma situations, a cross-table lateral can be substituted	MSP parallel to the IR; IPL perpendicular to the IR; adjust neck so IPL is perpendicular to edge of cassette; in trauma situations, use a vertical grid device and slightly elevate the head off of the table	Perpendicular 2 inches superior to EAM	Lateral image of cranium with increased detail of side closest to IR, sella turcica, dorsum sellae
Skull—PA Axial Projection (Caldwell Method)	Upright or prone; MSP to ML of the IR; forehead and nose touching grid device	OML perpendicular to IR; MSP perpendicular to IR; center IR at nasion	15 degree caudal angle directed through nasion	Frontal bone, frontal sinuses, facial bones, crista galli, petrous ridges projected in the lower third of the orbits
Skull—PA Projection	Upright or prone; MSP to ML of the IR; forehead and nose touching grid device	OML perpendicular to IR; MSP perpendicular to IR; center IR at nasion	Perpendicular directed through nasion	Frontal bone, petrous ridges filling the orbits completely
Skull—AP Projection or AP Axial Projection	Upright or supine; depress chin until OML is perpendicular to the IR	OML perpendicular to IR; MSP perpendicular to IR; center IR at nasion	Perpendicular directed through nasion or 15 degree cephalic angle through nasion	Frontal bone; orbits magnified due to increased OID
Skull—AP Axial Projection (Towne Method)	Seated upright or supine; MSP centered to ML of grid device	OML perpendicular to IR; if patient cannot flex neck sufficiently, then Infraorbital-Meatal line (IOML) can be used perpendicular to IR; upper margin of the IR adjusted to level of cranial vertex	If OML is used, 30 degree caudal angle; if Infraorbital-Meatal line (IOML) is used, 37 degree angle CR enters approximately 2.5 inches above the glabella and passes through the level of EAM	Posterior parietals, occipital bone, dorsum sellae through the foramen magnum

	Position of Patient	Position of Part	Central Ray	Structures
Skull—PA Axial Projection (Haas Projection)	Upright or prone; MSP of patient to ML of grid device; forehead and nose on table	OML perpendicular to IR	25 degree cephalic angle entering 1.5 inches below inion, exiting 1.5 inches superior to nasion	Occipital bone
Submentovertical Projection	Seated upright; place the IOML parallel to the IR by tilting the head back	MSP of body to ML of grid device; IOML parallel to the cassette; MSP of head perpendicular to the cassette	Perpendicular to IOML; entering MSP of the throat between the mandible and passes a point three-quarters of an inch anterior to the level of the EAM	Cranial base including sphenoidal and ethmoidal sinuses
Facial Bones— Lateral Projection	Patient upright or semi-prone; in trauma situations, a cross-table lateral can be substituted	MSP parallel to the IR; IPL perpendicular to the IR; adjust neck so IPML is perpendicular to edge of cassette; in trauma situations, use a vertical grid device and slightly elevate the head off of the table	Perpendicular entering the lateral surface of the zygoma halfway between outer canthus and EAM	All facial bones
Parietoacanthial Projection (Waters Method)	Patient either upright or prone; MSP of patient to ML of grid device	Rest extended chin on grid device, hyperextend the chin until OML forms a 37 degree angle to the plane of the IR; MSP of head perpendicular to IR; center IR to level of acanthion	Perpendicular to exit the acanthion	Orbits, maxilla, and zygomatic arches
Parietoacanthial Projection (Modified Waters Method)	Patient either upright or prone; MSP of patient to ML of grid device	Rest extended chin on grid device, hyperextend the chin until OML forms a 55 degree angle to the plane of the IR; MSP of head perpendicular to IR; center IR to level of acanthion	Perpendicular to exit the acanthion	Facial bones with less angulation, petrous ridges are below the orbits

	Position of Patient	Position of Part	Central Ray	Structures
Nasal Bones—Lateral Projection	Patient in a semi-prone position; MSP of head is horizontal	MSP parallel to IR; IPL perpendicular to IR; flex neck so IOML is parallel to transverse axis of IR	Perpendicular to nose half an inch distal to nasion	Nasal bones
Zygomatic Arches—Submentovertical Projection	Seated upright; place the IOML parallel to the IR by tilting the head back	MSP of body to ML of grid device; IOML parallel to the cassette; MSP of head perpendicular to the cassette	Perpendicular to IOML approximately 1 inch posterior to outer canthi	Bilateral symmetric zygomatic arches free of superimposition
Zygomatic Arches—Tangential Projection (May Method)	Seated upright or prone; place the IOML parallel to the IR by extending the head back	MSP rotated 15 degrees away from affected side; IOML parallel to the IR; tilt top of head away from affected side approximately 15 degrees	Perpendicular to IOML through zygomatic arch, approximately 1.5 inches posterior to outer canthus	Unilateral zygomatic arch free of superimposition
Mandible—Axiolateraloblique Projection	Semi-prone or semi-supine position; head in lateral position; IPL perpendicular to the cassette; extend patient's neck so that the mandible is parallel to the IR	Adjust rotation of head parallel to cassette	25 degree cephalic angle to pass through the mandibular region of interest	Mandible that is parallel to the cassette
TMJ—AP Axial Projection	Supine or upright	Adjust head so MSP is perpendicular to the IR; flex neck so OML is perpendicular to IR	35 degree caudal angle centered at TMJ 3 inches above the nasion	Condyles of mandible and TMJ
TMJ—Axiolateral Projection (Open and Closed Mouth)	Upright or semi-prone	Center the IR to a point 1.5 inches anterior to EAM; IPL perpendicular to IR	25–30 degree caudal angle entering 1.5 inches anterior and 2 inches superior to upside EAM	TMJ with mouth open and closed

	Position of Patient	Position of Part	Central Ray	Structures
Sinuses—Lateral Projection	Patient upright or semi-prone; in trauma situations, a cross-table lateral can be substituted	MSP parallel to the IR; IPL perpendicular to the IR; adjust neck so IPML is perpendicular to edge of cassette	Perpendicular entering 0.5–1 inch posterior to outer canthus	All four sinus groups
Sinuses— PA Projection (Caldwell Method)	Same as above for skull	Same as above	Same as above	Frontal sinus, sphenoidal sinus through nasal fossa
Sinus— Parietoacanthial Projection (Waters Method)	Same as above for facial bones	Can be performed with open mouth modification		Maxillary sinus with petrous ridges below; the sphenoid sinuses are now visible through the open mouth
Sinus—SMV	Seated upright; place the IOML parallel to the IR by tilting the head back	MSP of body to ML of grid device; IOML parallel to the cassette; MSP of head perpendicular to the cassette	Perpendicular to IOML; enters MSP of the throat between the mandible and passes a point three-quarters of an inch anterior to the level of the EAM	Sphenoid and ethmoid sinus
Orbits— Parieto-Orbital Projection (Rheese Method)	Semi-prone or upright	Affected orbit centered to IR; rest zygoma, nose, and chin on grid device; adjust flexion so AML perpendicular to the cassette; MSP forms a 53 degree angle to the plane of the cassette	Perpendicular entering 1 inch superior and posterior to upside ear attachment, exiting orbit closest	Optic canal on end and optic foramen on the inferior lateral quadrant of the affected orbit

Chapter 8 Review Questions

1. The _____ is the largest carpal bone.
   a. navicular
   b. pisiform
   c. tapezoid
   d. capitate

2. Localization of the hip is obtained by bisecting a line between
   a. the symphysis pubis and iliac crest.
   b. the iliac crest and ASIS.
   c. the symphysis pubis and ASIS.
   d. the iliac crest and pelvic inlet.

3. The position best used to demonstrate all four paranasal sinuses is
   a. the parietoacanthial projection (Waters Method).
   b. the AP axial projection (Caldwell Method).
   c. the submentovertex.
   d. the lateral.

4. The best position to use in order to rule out a fracture of the base of the fifth metatarsal is
   a. the medial oblique projection.
   b. the lateral projection.
   c. the dorsoplantar projection.
   d. the plantodorsal projection.

5. The posterior oblique projection (Grashey Method) is used to best demonstrate which area of the shoulder girdle?
   a. scapula
   b. AC joint
   c. glenoid fossa
   d. humeral head

6. The patella is classified as
   a. a sesamoid.
   b. a long bone.
   c. an irregular bone.
   d. a flat bone.

7. In order to best visualize the intervertebral space between C-7 and T-1, especially in patients who cannot fully depress their shoulders, the position best used is the
   a. AP.
   b. swimmer's.
   c. anterior oblique.
   d. posterior oblique.

8. A PA chest projection is performed at
   a. 36 inches.
   b. 40 inches.
   c. 48 inches.
   d. 72 inches.

9. The projection that best demonstrates the occipital bone of the cranium is the
   a. AP.
   b. lateral.
   c. AP axial Townes.
   d. submentovertical.

10. In the lateral skull projection, the baseline which is perpendicular to the plane of the IR is
    a. OML.
    b. IPL.
    c. IOML.
    d. GML.

11. The right sacroiliac joint would be best demonstrated in which projection?
    a. lateral projection
    b. RPO projection
    c. RAO projection
    d. AP projection

12. What type of curvature does the thoracic spine present?
    a. lordotic
    b. kyphotic
    c. linear
    d. transverse

13. A postoperative contrast procedure in which contrast is injected into the common bile duct via an external catheter is called
    a. an upper GI series.
    b. a barium enema.
    c. an ERCP.
    d. a T-tube cholangiography.

14. In a projection of the shoulder in which the patient is in an AP position with the body rotated 45 degrees toward the shoulder of interest, the part being demonstrated is the
    a. glenoid fossa.
    b. body of the scapula.
    c. AC joint.
    d. greater tuberosity.

15. Prior to a barium enema, the catheter tip should be inserted into the rectum no more than
    a. 2–3 inches.
    b. 6 inches.
    c. 1 inch.
    d. 3.5–4 inches.

16. If the navicular bone and its articulations are of interest when radiographing the wrist, what position is best?
    a. lateral
    b. Stecher
    c. carpal canal
    d. AP

17. The aspect of the lung that lies superior to the clavicles is called the
    a. cardiophrenic angle.
    b. carina.
    c. costophrenic angle.
    d. apex.

18. The position of the elbow which would best demonstrate the radial head, neck, and tuberosity free of superimposition is the
    a. AP.
    b. medial oblique.
    c. lateral oblique.
    d. lateral.

19. Which position is best in order to demonstrate the zygapophyseal articulations of the lumbar spine?
    a. AP projection
    b. AP 45 degree oblique projection
    c. lateral projection
    d. AP erect projection

20. When imaging the orbit utilizing the Rheese Method, if the proper position is utilized, the optic foramen should be demonstrated in which quadrant?
    a. upper lateral quadrant
    b. upper medial quadrant
    c. lower lateral quadrant
    d. lower medial quadrant

21. The projection of the paranasal sinuses that best demonstrates the maxillary sinus is the
    a. lateral projection.
    b. AP axial projection (Caldwell Method).
    c. parietoacanthial projection (Waters Method).
    d. submentovertical projection.

22. In order to demonstrate the carpal pisiform free of superimposition, the position best used is the
    a. PA projection.
    b. PA oblique projection.
    c. AP oblique projection.
    d. lateral projection.

23. The largest bone in the lower leg is the
    a. tibia.
    b. fibula.
    c. femur.
    d. calcaneous.

24. The three aspects of the small intestine, in order, starting from the stomach are the
    a. ilium, jejunum, and duodenum.
    b. jejunum, ilium, and duodenum.
    c. duodenum, ilium, and jejunum.
    d. duodenum, jejunum, and ilium.

25. Which of the following is not located on the distal row of carpals?
    a. Hamate
    b. Capitate
    c. Navicular
    d. Trapezium

26. In order to best demonstrate possible fractures of the fifth metatarsal base, the projection utilized is the
    a. plantodorsal projection.
    b. AP lateral oblique.
    c. lateral.
    d. AP medial oblique.

27. The Fuchs Method demonstrates the odontoid process through the
    a. foramen magnum.
    b. sphenoid sinus.
    c. obturator foramen.
    d. open mouth.

28. The plane of the body which separates the body into equal right and left halves is the
    a. midcoronal plane.
    b. midsagittal plane.
    c. transverse plane.
    d. oblique plane.

29. When bilateral hip fractures are suspected, the radiographer should utilize which of the following methods in order to image the hip in a lateral position?
    a. Cleaves Method
    b. modified Cleaves Method
    c. Judet Method
    d. Clements-Nakayama Method

30. In order to best demonstrate the lower ribs, when should the exposure be made?
    a. after the patient takes a full deep breath and holds it
    b. on the second full inspiration of the patient
    c. on the patient's full exhalation
    d. while the patient is utilizing a breathing technique

31. When performing a lateral image of the sternum, where should the patient keep his or her hands?
    a. at his or her sides
    b. clasped behind the back
    c. directly out in front of the body
    d. on top of the head

32. The most frequently fractured carpal bone is the
    a. scaphoid.
    b. triquetrum.
    c. hamate.
    d. capitate.

33. The tuberosity of a rib articulates with which portion of a thoracic vertebra?
 a. body
 b. transverse process
 c. pedicle
 d. laminae

34. Which part of the mandible articulates with the temporal bone of the cranium?
 a. angle
 b. body
 c. condylar process
 d. coronoid process

35. When an image of the shoulder presents the lesser tubercle in profile medially, what position was the patient in?
 a. AP neutral
 b. AP internal rotation
 c. AP external rotation
 d. AP oblique

36. Which radiographic baseline is perpendicular to the image receptor in a lateral projection of the facial bones?
 a. interpupillary line
 b. infraorbitomeatal line
 c. orbitomeatal line
 d. acanthiomeatal line

37. Where should the central ray enter for a PA projection of the hand?
 a. midcarpal area
 b. third interphalangeal joint
 c. base of the third metacarpal
 d. base of the second metacarpal

38. For which of the following projections would a breathing technique be most useful?
 1. lateral thoracic spine
 2. transthoracic Lawrence Method
 3. AP pelvis

 a. 1 only
 b. 2 only
 c. 1 and 2 only
 d. 1, 2, and 3

39. Which projection of the ankle is used in order to best demonstrate the ankle mortise?
 a. AP
 b. AP medial oblique 15–20 degrees
 c. AP medial oblique 45 degrees
 d. lateral

40. An image is presented showing the stomach with barium within the fundus while the body and pylorus are filled with air. The position demonstrated is the
 a. AP oblique LPO.
 b. PA.
 c. PA oblique RAO.
 d. PA Trendelenburg.

41. Which position of the foot best demonstrates the sinus tarsi?
 a. AP lateral oblique projection
 b. lateral projection
 c. dorsoplantar projection
 d. AP medial oblique projection

42. If the patient is placed in an LAO position with the pelvis rotated 25 degrees and the center ray exiting 1 inch medial to the elevated ASIS, what is the structure shown?
 a. right sacroiliac joint
 b. left sacroiliac joint
 c. left ischial rami
 d. right ischial rami

43. For visualization of the occipital bone, if a patient is upright, up against the upright wall device with the midsagittal plane of the head and the baseline IOML perpendicular to the image, where is the central ray?
  a. 20 degrees caudad
  b. 30 degrees caudad
  c. 37 degrees caudad
  d. 45 degrees caudad

44. The flexure where the transverse colon meets the descending colon is called the
  a. hepatic flexure.
  b. hiatus flexure.
  c. gastric flexure.
  d. splenic flexure.

45. Which vertebral level does the iliac crest correspond to?
  a. L-3
  b. T-7
  c. T-4
  d. L-5/S-1

46. The part of a typical vertebra which is short and connects the body to the laminae on either side is called
  a. the transverse process.
  b. the pedicle.
  c. the spinous process.
  d. the inferior articulating process.

47. Which of the following projections of the chest would best demonstrate a right-sided pneumothorax?
  1. right lateral decubitis
  2. left lateral decubitis
  3. upright
  a. 1 only
  b. 2 only
  c. 2 and 3 only
  d. 1, 2, or 3

48. In a lateral projection of the knee, the central ray is angled 5 degrees cephalic. What is the purpose of this?
  a. to minimize the magnification of the patella
  b. to show small fractures of the tibial eminence
  c. to prevent superimposition of the medial condyle in the joint space
  d. to bring the fibula on the same plane as the rest of the leg

49. It is absolutely necessary when positioning a patient for sinus studies that all four projections are taken with the patient
  a. upright.
  b. supine.
  c. in the Trendelenberg position.
  d. prone.

50. In the PA oblique projection of the wrist, the carpals best visualized are the
  a. pisiform and hamate.
  b. greater multangular and lesser multangular.
  c. scaphoid and trapezoid.
  d. capitate and lunate.

51. What is the bony protuberance that can be palpated on the anterior tibia?
  a. tibial spine
  b. tibial eminence
  c. tibial plateau
  d. tibial tuberosity

52. The baseline which is an imaginary line drawn from the external auditory meatus to the outer canthus of the eye is called the
  a. interpupillary line.
  b. orbitomeatal line.
  c. ancanthiomeatal line.
  d. mentomeatal line.

53. In an AP projection of the coccyx, what is the degree of central ray angulation?
a. 10 degrees cephalic
b. 15 degrees caudal
c. 15 degrees cephalic
d. 10 degrees caudal

54. During imaging of the shoulder, which view best demonstrates the greater tubercle in profile?
a. AP projection (external rotation)
b. AP projection (internal rotation)
c. PA axial oblique, supraspinatus outlet projection
d. axial projection

55. In which of the following organs is peristalsis considered strongest?
a. esophagus
b. stomach
c. small intestine
d. colon

56. The projection for best visualizing the proximal tibiofibular joint is the
a. anteroposterior projection of the knee.
b. anteroposterior oblique (10 degrees) projection of the ankle.
c. anteroposterior oblique projection with internal rotation of the knee.
d. anteroposterior oblique projection with external rotation of the knee.

57. For a parietoacanthial projection (Waters Method) of the sinuses, the orbitomeatal line should form a ___ angle to the plane of the IR.
a. 25 degree
b. 30 degree
c. 37 degree
d. 40 degree

58. For a lateral projection of the knee, what is the degree of flexion required?
a. 10–20 degrees
b. 20–30 degrees
c. 30–40 degrees
d. 40–50 degrees

59. Which of the following anatomical structures is (are) located in the left upper quadrant?

    1. stomach
    2. liver
    3. spleen

a. 1 only
b. 2 only
c. 1 and 3 only
d. 1, 2, and 3

60. Which of the following is NOT a cranial bone?
a. vomer
b. occipital
c. parietal
d. temporal

61. In an examination of the thoracic spine, the intervertebral foramina is best visualized in which view?
a. AP
b. PA oblique RAO/LAO
c. AP oblique RPO/LPO
d. lateral

62. The term *axis* best refers to
a. C-1.
b. C-2.
c. C-7.
d. L-1.

63. The articulations between the zygapophyseal joints of the cervical vertebrae are best seen in which projection?
    a. anteroposterior
    b. lateral
    c. open mouth
    d. PA oblique

64. Which of the following projections best demonstrate(s) the intercondylar fossa of the knee?

    1. Holmblad
    2. Camp Coventry
    3. Danelius-Miller

    a. 1 only
    b. 2 only
    c. 1 and 2 only
    d. 1, 2, and 3

65. When performing a barium enema, if the splenic flexure of the colon is of interest, then the projection that will best demonstrate this structure is the
    a. PA.
    b. LPO.
    c. RPO.
    d. AP axial.

66. What is the degree of obliquity for a medial oblique of the foot?
    a. 20 degrees
    b. 30 degrees
    c. 45 degrees
    d. 69 degrees

67. For an AP projection of the sacrum, the central ray is angled
    a. 15 degrees cephalic.
    b. 15 degrees caudal.
    c. 10 degrees cephalic.
    d. 10 degrees caudal.

68. The first seven ribs are classified as
    a. floating ribs.
    b. false ribs.
    c. cervical ribs.
    d. true ribs.

69. On C-2, the vertical process that extends upward and articulates with C-1 is called the
    a. styloid process.
    b. coronoid.
    c. odontoid.
    d. superior articulating process.

70. To ensure the proper preparation of a patient prior to performing a double contrast barium study of the stomach, the patient is required to be NPO for a minimum duration of
    a. 1 hour.
    b. 4 hours.
    c. 8 hours.
    d. 24 hours.

71. The number of posterior ribs visualized in a properly exposed posteroanterior chest x-ray is
    a. 6.
    b. 8.
    c. 10.
    d. 12.

72. The sthenic body habitus makes up what percentage of the population?
    a. 5%
    b. 35%
    c. 10%
    d. 50%

73. In a case where there is a suspected hip fracture, which view(s) should be performed?

    1. AP

    2. Modified Cleaves

    3. Danelius-Miller

a. 1 only

b. 1 and 2 only

c. 2 and 3 only

d. 1 and 3 only

74. When there is a suspected fracture of the right axillary ribs, the best projection to be performed is the

a. LAO.

b. LPO.

c. PA.

d. AP.

75. The absorption of nutrients primarily takes place in which organ?

a. liver

b. small intestine

c. stomach

d. colon

76. During a barium enema study, in order to facilitate the proper flow of contrast into the colon, the enema bag should be located

a. 18–24 inches above the anus.

b. 10–17 inches above the anus.

c. 3.5–4 inches above the anus.

d. 12–18 inches below the anus.

77. In what projection is a patient semi-prone with the head in a lateral position, with the central ray directed 25 degrees cephalic passing through the region of interest?

a. Haas Method

b. Townes Method

c. axiolateral oblique method

d. Law Method

78. A boxer's fracture best describes a fracture of the

a. metacarpals.

b. metatarsals.

c. phalanges.

d. navicular.

79. One primary function of the liver is to create _____ to aid in digestion.

a. chyme

b. feces

c. bile

d. mucous

80. The largest long bone in the body is the

a. tibia.

b. humerus.

c. ulna.

d. femur.

81. The _____ skull forms an angle of 54 degrees between the petrous pyramids and the midsagittal plane.

a. brachiocephalic

b. dolichocephalic

c. brachycephalic

d. mesocephalic

82. When visualizing the dens through the open mouth, which two structures need to be aligned in the same vertical plane?

a. mastoid tip and hard palate

b. mastoid tip and lower edge of the upper incisors

c. EAM and the lower edge of the upper incisors

d. EAM and the nasal septum

83. The most lateral aspects of the lungs in the PA projection are called
  a. the apices.
  b. the cardiophrenic angles.
  c. the costophrenic angles.
  d. the hilum.

84. Urine that is produced in the kidney is transported to the bladder via the
  a. ureter.
  b. urethra.
  c. aorta.
  d. inferior vena cava.

85. The common bile duct connects to which of the following aspects of the digestive tract?
  a. cecum
  b. duodenum
  c. jejunum
  d. ilium

86. The vertebral column is composed of ___ vertebrae.
  a. 23
  b. 33
  c. 37
  d. 41

87. The aspect of the elbow best seen in profile in an AP medial oblique projection is the
  a. coracoid.
  b. olecranon.
  c. radial head.
  d. coronoid.

88. When a posterior or anterior dislocation of the humeral head is in question, the degree of obliquity required in this projection is
  a. 30 degrees.
  b. 40 degrees.
  c. 60 degrees.
  d. 90 degrees.

89. How many bones are in the normal human body?
  a. 185
  b. 200
  c. 206
  d. 212

90. Common fractures of the shoulder take place at which aspect of the humerus?
  a. surgical neck
  b. anatomical neck
  c. humeral head
  d. glenoid fossa

91. In the PA oblique projection of the hand, what is the position of the thumb?
  a. PA
  b. AP
  c. oblique
  d. lateral

92. The patellofemoral joint is best visualized free of the femur in which of the following methods?
  a. Grashey
  b. Settegast
  c. Judet
  d. Lawrence

93. What does the acronym ERCP stand for?
  a. esophageal retro calcaneal projection
  b. endoscopic retrograde cholangiogram procedure
  c. esophageal retrograde pancreatic procedure
  d. endoscopic retrograde cholangiopancreatography

94. Which plane of the body separates the body into anterior and posterior portions?
  a. sagittal
  b. coronal
  c. transverse
  d. oblique

95. The external portion of the ear is commonly called the
a. auricle.
b. EAM.
c. petrous portion.
d. mastoid process.

96. If a patient is placed recumbent with the feet elevated in relation to the rest of the body, this position is commonly referred to as
a. Fowler's.
b. decubitis.
c. Trendelenberg.
d. prone.

97. In positioning the abdomen, which projection(s) best demonstrate(s) free air?

    1. KUB
    2. left lateral decubitis
    3. upright

a. 1 only
b. 1 and 2 only
c. 2 and 3 only
d. 1, 2, and 3

98. While in the anatomical position, the palmar portions of the hands are in the
a. supinated position.
b. pronated position.
c. inverted position.
d. adducted position.

99. Another common name for the external occipital protuberance is
a. gonion.
b. acanthion.
c. inion.
d. glabella.

100. Which structure separates the thorax from the abdominal cavities?
a. aorta
b. diaphragm
c. hilum
d. mediastinum

CHAPTER 8 ANSWERS

1. **d.** The capitate is the largest carpal bone.
2. **c.** A line is bisected between ASIS and symphysis pubis and dropped inferiorly and perpendicularly 1 inch for hip localization.
3. **d.** The lateral projection demonstrates all four paranasal sinuses.
4. **a.** The 30 degree medial oblique of the foot best demonstrates the fifth metatarsal tuberosity.
5. **c.** The 35–45 degree posterior oblique of the shoulder (Grashey Method) is used to demonstrate the glenoid cavity.
6. **a.** The patella is the largest sesamoid in the body.
7. **b.** The swimmer's position is best used for trauma/broad shouldered patients in order to best visualize the lower cervical/upper thoracic vertebrae.
8. **d.** Chest projections are performed at 72 inches in order to minimize heart magnification.
9. **c.** The AP axial Townes Projection best demonstrates the occipital bone due to its 30 degree caudal angle, which frees it from superimposition.
10. **b.** IPL is perpendicular to the plane of the IR in a lateral projection of the cranium. IOML is parallel to the plane of the IR.
11. **c.** The right SI joint is best demonstrated utilizing the RAO (or LPO) projection.
12. **b.** The T-spine has a kyphotic curve.
13. **d.** T-tube cholangiography is performed postoperatively via contrast injection into a T-tube catheter inserted into the common bile duct for patency.
14. **a.** The AP 45 degree oblique projection of the shoulder (or Grashey Method) is used to best demonstrate glenoid fossa.
15. **d.** The catheter should be inserted into the rectum no more than 3.5–4 inches.
16. **b.** The Stecher view, which is a PA projection in ulnar flexion with the fingers elevated 20 degrees from the plane of the IR, is best to visualize the navicular.
17. **d.** The apex is the area of the lung lying over the clavicles.
18. **c.** The lateral 45 degree oblique of the elbow demonstrates the proximal radius free of superimposition.
19. **b.** The AP oblique projections (LPO/RPO) of the lumbar spine are used to best demonstrate the zygapophyseal articulations.
20. **c.** In a properly positioned Rheese Projection, the optic foramen is visualized in the lower lateral quadrant of the orbit.
21. **c.** For maxillary sinuses, the Waters Projection is best. It is a PA projection utilizing a 37 degree angle between the tip of the nose and the IR with the chin resting on the upright grid device with the CR exiting the acanthion.
22. **c.** The AP oblique projection in lateral rotation, with the wrist laterally rotated 45 degrees from the AP, demonstrates the carpal pisiform free of superimposition from the triquetrum.
23. **a.** The tibia is the longest bone in the lower leg.
24. **d.** The order in which the small intestine originates from the stomach is duodenum, jejunum, and ilium.
25. **c.** The navicular is located in the proximal row of carpals.
26. **d.** The AP medial oblique 30 degree internal rotation is used to demonstrate the lateral side of the foot, including the base of the fifth metatarsal.
27. **a.** The Fuchs Method projects the dens through the foramen magnum.
28. **b.** The midsagittal plane separates the body into equal right and left halves.
29. **d.** The Clements-Nakayama Method is best utilized in order to image the hip and proximal femur if bilateral hip fractures are suspected,

as it handles both the center ray and the film posteriorly to the femoral neck in order to visualize the hip in a lateral position.

30. **c.** For the lower ribs, the exposure should be made after exhalation in order to utilize the diaphragm as contrast.

31. **b.** For a lateral sternum, the hands should be firmly clasped behind the back in order to help the patient push out his or her chest, thereby freeing the sternum from most superimposition.

32. **a.** The scaphoid (navicular) is the most frequently fractured carpal.

33. **b.** The tuberosity of a rib articulates with the transverse process of a thoracic vertebrae.

34. **c.** The condylar process articulates with the temporal bone at the TMJ.

35. **b.** The AP internal rotation demonstrates the lesser tubercle of the humeral head in profile medially.

36. **a.** IPL is perpendicular to the film in a lateral projection of the facial bones.

37. **c.** The CR of the PA hand enters the base of the third metacarpal.

38. **c.** The lateral thoracic spine and the transthoracic humerus can both utilize the breathing technique in order to blur out the structures of the lungs lying above (and below) the areas of interest.

39. **b.** The ankle mortise requires a 15–20 degree internal rotation of the leg with a central ray perpendicular to the ankle joint.

40. **a.** The AP oblique LPO position will demonstrate the fundus and duodenal bulb opacified by positive contrast media while the pylorus and body are filled with air.

41. **d.** The AP medial oblique best demonstrates the sinus tarsi, the cuboid, and the lateral aspects of the foot.

42. **b.** In the LAO or RPO positions for SI joints, the left SI joint is the part of interest.

43. **c.** If IOML is used for the AP axial Townes Method for cranium, a 37 degree central ray is utilized; for OML it is 30 degrees.

44. **d.** The splenic flexure is located at the junction between the transverse and descending colon.

45. **a.** The iliac crest corresponds to the vertebral level of L-3.

46. **b.** The pedicles emerge from the posterior portion of the body on each side and lead to the laminae.

47. **c.** In order to demonstrate a right-sided pneumothorax on a patient, that patient should either be positioned upright or placed in a left lateral decubitis because free air will rise in the chest.

48. **c.** The central ray of a lateral knee is angled 5 degrees cephalic in order to prevent the superimposition of the medial femoral condyle, which naturally is larger and lower in a patient's body.

49. **a.** It is imperative that all sinus studies be performed erect in order to properly differentiate air/fluid levels within the paranasal sinuses.

50. **c.** The PA oblique projection of the wrist best demonstrates the lateral carpal bones: the scaphoid, the trapezoid, and the trapezium.

51. **c.** The tibial plateau is palpable on the anterior proximal tibia.

52. **b.** The orbitomeatal line refers to the baseline drawn between the outer canthus of the eye and the EAM and is utilized in positioning of the skull.

53. **d.** For an AP projection of the coccyx, the central ray is 10 degrees caudal entering MSP 2 inches inferior to the level of ASIS.

54. **a.** The greater tubercle is in profile on the lateral aspect of the humerus during an AP projection when the arm is in an external rotation.

55. **a.** Peristalsis is strongest in the beginning of the digestive process, which starts after swallowing

in the esophagus and proceeds through the stomach to small bowel and finally the colon where peristalsis is the slowest.

56. c. In order to best visualize the distal tib/fib joint, the knee should be internally rotated 45 degrees with the central ray directed half an inch below the apex of the patella.

57. c. For a parietoacanthial projection (Waters Method) of the sinuses, the orbitomeatal line should form a 37 degree angle to the plane of the IR. This is done by tilting the head back until this angle is attained.

58. b. In order to relax the muscles of the knee when performing a lateral view, it is best to have the patient flex the affected knee 20–30 degrees.

59. c. The stomach and spleen are located in the left upper quadrant of the abdomen.

60. a. The vomer is one of the 14 facial bones, not a cranial bone.

61. d. The intervertebral foramina is best visualized in the lateral view for the thoracic spine.

62. b. Axis refers to C-2, which allows the head to pivot left and right.

63. b. The zygapophyseal joint is best visualized in the lateral view of the cervical spine.

64. c. The PA axial Camp Coventry Method and the AP axial Holmblad Method of the knee both allow for visualization of the open intercondylar fossa.

65. b. The LPO projection of a barium enema series will allow the splenic flexure to be visualized with limited superimposition of the adjacent areas of the colon.

66. b. The medial oblique of the foot requires the foot to be rotated medially 30 degrees. Any further obliquity would superimpose the medial metatarsals over the lateral.

67. a. For an AP projection of the sacrum, the central ray is angled 15 degrees cephalic directed 2.5 inches superior to symphysis pubis.

68. d. The first seven ribs are classified as true ribs due to their direct articulation with the sternum.

69. c. The odontoid process, or dens, extends vertically from the body of C-2 and articulates with C-1. It helps to provide the pivot for the head.

70. c. Eight hours is the minimal time necessary in order for the stomach to be empty of all contents prior to an upper GI series.

71. c. There should be 10 posterior ribs visualized in a properly exposed PA chest x-ray done upon the second full inspiration.

72. d. Sthenic patients make up 50% of the population.

73. d. For trauma hip radiography, the technologist should perform an AP and a cross-table lateral (Danelius-Miller Method). The affected leg should never be abducted, which is essential when positioning for the modified Cleaves Method.

74. a. The LAO position for ribs best demonstrates the right axillary ribs, which is the side farthest from the IR. The RPO position also demonstrates the right axillary ribs.

75. b. The small intestine is the primary site of nutrient absorption.

76. a. The enema bag should be placed 18–24 inches above the anus in order to facilitate the flow of barium and to best opacify the colon.

77. c. The axiolateral oblique projection for mandibular ramus requires the head to be in a true lateral position with the central ray directed 25 degrees cephalic.

78. a. A boxer's fracture is a fracture of the metacarpals, generally after impact of a punch.

79. c. Bile is created in the liver.

80. d. The femur is the largest bone of the body.

81. c. The brachycephalic skull forms an angle of 54 degrees from the petrous pyramids and the MSP. It is short from front to back, broad from side to side, and shallow from vertex to base.

82. b. The mastoid tip and the lower edge of the upper incisors need to be perpendicular to the film in order for the dens to be visualized through the open mouth.

83. c. The costophrenic angles are the most lateral aspect of the lungs in a PA view.

84. a. The ureters are the connection from the kidneys to the bladder. Urine is transported down.

85. b. The common bile duct empties contents into the duodenum via the ampulla of vater.

86. b. There are 33 vertebrae in the spinal column (7 cervical, 12 thoracic, 5 lumbar, 5 fused sacral, 4 fused coccyx).

87. d. The coronoid process is seen in profile in an AP medial oblique projection of the elbow.

88. c. In the PA scapular "Y" projection, the degree of obliquity is 60 degrees.

89. c. There are 206 bones in the body.

90. a. The most frequently fractured site of the shoulder is the surgical neck, which is a very thinned out section of the humerus.

91. c. The thumb lies naturally on an oblique plane in a PA projection of the hand.

92. b. The Settegast (or sunrise) projection best demonstrates the patella free of superimposition as well as the patellofemoral joint.

93. d. It stands for endoscopic retrograde cholangiopancreatography.

94. b. The coronal plane refers to a plane that separates the structure into anterior and posterior portions.

95. a. The external part of the ear is referred to as the auricle.

96. c. Trendelenberg is when a patient is recumbent with his or her feet elevated above the head.

97. c. The best projections for demonstrating free air are left lateral decubitus with the patient lying on the left side with a horizontal center ray at the level of iliac crest and the upright projection with the upper margin of the IR above the level of the diaphragm.

98. a. In the anatomical position, the arms are slightly abducted with the palms in a supinated position with feet slightly spread.

99. d. The inion is commonly referred to as the external occipital protuberance.

100. b. The diaphragm represents the border between the thorax and abdomen.

REVIEW OF PATIENT CARE AND EDUCATION

Key Terms

advance directive	battery
airborne precautions	blood pressure
airborne transmission	bradycardia
allergic shock (anaphylaxis)	cardiac arrest
anaphylactic reactions	cardiogenic shock
aqueous iodine compound	cardiovascular reactions
assault	chest tube

Key Terms

common vehicle transmission	medical asepsis
contact precautions	nasogastric (NG) tubes
diastolic pressure	negative contrast agent
direct contact transmission	negligence
diversity	neurogenic shock
do not resuscitate order (DNR)	nonverbal communication
droplet precautions	oxygen administration
droplet transmission	patient bill of rights
durable power of attorney	patient education
false imprisonment	patient history
hand washing	positive contrast agent
HIPAA	pulse
hypodermic needle gauge	res ipsa loquitur
hypovolemic shock	respiration
implied consent	respiratory arrest
indirect contact transmission	respondeat superior
infection control barriers	septic shock
informed consent	shock
invasion of privacy	slander
iodinated ionic contrast agents	sphygmomanometer
iodinated nonionic contrast agents	standard precautions
IV catheter	suction unit
libel	surgical asepsis
material safety data sheets (MSDs)	systolic pressure

Key Terms

tachycardia	vector-borne transmission
temperature	ventilators
torts	verbal communication
trauma	

Medicolegal Aspects of Practice

Torts

- Violations of civil law
- Commonly referred to as personal injury law
- Injured parties have a right to compensation

Intentional Misconduct

- **Assault:** The threat of injurious touching. An assault can be justified on the patient's behalf if he or she feels threatened or believes he or she will be touched injuriously.
- **Battery:** Unlawful touching of a person without his/her consent. A patient's refusal to be touched must be respected. Performing radiographs on the wrong patient may be considered battery.
- **False imprisonment:** A detention of a person against his or her will. Use of restraints is not justified unless there is a question about the safety of the patient.

Invasion of Privacy

- violation of HIPAA
- unnecessarily or improperly touching or exposing the patient's body without his or her permission

- photographing patients without their permission or written consent
- libel, which is written documentation that results in defamation of character or loss of reputation
- slander, which is a malicious, false, and defamatory statement or report; verbally falsifying information that results in the spreading of rumors

Unintentional Misconduct (Negligence)

- **Neglect:** Omission of reasonable care or caution.
- **Gross negligence:** Conscious and voluntary; disregard of the need to use reasonable care; performance of acts that demonstrate disregard for life or limb.
- **Contributory negligence:** Injured person is a contributing party to the injury.

Four Conditions Needed to Establish Malpractice

1. Establishment of standard of care
2. Demonstration that standard of care was violated by the named party
3. Demonstration that loss or injury was caused by named party
4. Demonstration that loss or injury truly occurred as a result of negligence

Legal Doctrines

- *Respondeat superior*—**Latin for "let the master answer"**: This is a legal doctrine that states that an employer will be held liable for an employee's actions.
- **Rule of personal responsibility**: This rule states that individuals will be held responsible for their own actions.
- *Res ipsa loquitur*—**Latin for "the thing speaks for itself"**: This is a doctrine which states that the cause was obvious.

Charting

- The chart is a document that uses legible writing regarding patient's condition, refusal of procedure, reaction to contrast, type, amount, and rate of contrast material injected, etc.
- Information must be time stamped with date, time, and initials of the radiographer.
- In the case of a mistake in documentation, draw a single line through the error, initial it, and rewrite the correct information.

Radiographs

Radiographs must contain the following information:

- name
- medical record number/patient identifier
- date
- left or right markers

Always keep in mind that a radiograph is a legal document that may be used in a court of law. Radiographs are personal information pertaining to the patient and must never be released without the appropriate consent.

Retention of radiographs:

- Radiographs are stored for seven years from the date of the examination.

- Mammograms are kept for the life of the patient.
- Radiographs of minors are retained for 18 or 21 years; this is dependent on the state of residence of the patient. Radiographs are maintained for seven years thereafter.

Patient's Rights

Bill of Rights

A patient's bill of rights is a statement of the rights to which a patient is entitled when receiving medical care. Typically, a statement describes the positive rights that doctors and hospitals should provide patients, thereby providing information, offering fair treatment, and granting them full autonomy over medical decisions.

Implied Consent

Implied consent provides for the care of the patient if he or she is in a state of unconsciousness. Implied consent is solely based on the assumption that the patient would approve care if he or she were conscious.

Informed Consent

A patient must be of legal age to certifiably sign or grant consent to perform a procedure. Any individual under the age of 18 is considered a minor and must have a parent or guardian sign on his or her behalf. As a medical professional, you should never force or insist that an individual consent to a procedure. Informed consent must be offered to you by the patient voluntarily of his or her own free will.

Health Insurance Portability and Accountability Act (HIPAA)

HIPAA was passed by the U.S. congress in 1996 to secure the rights of patient information only to those involved with the direct care of the patient. The patient has the right to prohibit the sharing of personal health information and must fully execute consent for the release of information to any third parties.

Do Not Resuscitate (DNR)

A DNR is a consented agreement between the patient or designee and physician that grants the patient the right to waive a CODE from being activated.

Advance Healthcare Directive

Advance healthcare directives, also known as living wills, are instructions given by the patient specifying what actions he or she wishes to have taken for his or her health in the event that the person is no longer able to make decisions due to illness or incapacity. The patient will also appoint a person to make such decisions on his or her behalf if the patient is unable to do so.

Power of Attorney

A power of attorney provides the right for another person to make decisions regarding medical care if the patient cannot communicate.

Communication with the Patient

Why Is It So Important to Demonstrate Clear and Open Communication?

- It establishes a trustworthy and professional relationship with your patient.
- It provides a detailed explanation of the examination.
- It demonstrates respect.
- It provides comfort for the patient if he or she needs to ask a question pertaining to the exam.

Verbal Communication

- When using verbal communication, you must speak clearly, slowly, and use a common vocabulary your patient can understand.
- Never show your frustrations or express them toward your patient if you are having a bad day. Verbal communication and the tone you use toward your patients can reflect negatively or positively upon you as a professional.

- Always remember to listen to what your patients have to say; be patient and calm.

Nonverbal Communication

- Use facial expressions.
- Use body motions.
- Be respectful of the patient's age and ethnicity in terms of eye contact. Certain ethnicities prohibit individuals from making eye contact.

Diversity in Medical Imaging

A radiographer must always be aware of the diversity of patients, which may include any of the following factors:

- age
- gender
- language
- race and ethnicity
- marital status
- political beliefs
- sexual preference
- socioeconomic background
- religious beliefs
- residence or geographic origin
- physical or mental disability

Other Departments

A well-rounded radiographer should be able to answer basic knowledge questions pertaining to other modalities. In the event that a patient must go for additional imaging or visit another department, he or she will assume you know the following modalities or where to find other departments such as:

- angiography
- bone densitometry
- computed tomography (CT)
- mammography
- magnetic resonance imaging (MRI)
- nuclear medicine

- positron emission tomography (PET)
- ultrasound
- admitting
- patient relations
- security
- rehabilitation
- social work
- patient financial services

Infection Control

Medical Asepsis

Medical asepsis is the effort to reduce the probability of infectious organisms being transmitted to a susceptible individual by reducing the total number of organisms through the following processes:

- **Cleanliness:** hand washing (for a minimum of 15 seconds).
- **Disinfection:** destruction of pathogens by using chemical materials.
- **Sterilization or surgical asepsis:** treating objects with heat, gas, or chemicals to make them free of germs.

Surgical Asepsis

Complete removal of all organisms from the equipment and environment, using the following techniques and materials:

- boiling item(s) for 12 minutes
- dry heat for 1–6 hours at temperatures of 300+ F
- gases used for electrical, plastic, and rubber items (freon and ethylene)
- steam
- chemicals
- ionizing radiation (commercial)
- nonionizing radiation (microwaves and low-pressure steam)

Sterile Techniques

- Sterile trays should be opened so that the first flap opens away from surgeon.
- Sterile solutions should be opened without touching the inside of the lid or opening of the container.
- Sterile gowns are not sterile in the back.
- Sterile gloves must be kept above the waist.
- Masks must fit snugly over the mouth and nose.
- When in the operating room or any sterile environment, always pass back to back.
- Any sterile item touched by a nonsterile person is no longer sterile and must be either disposed or sent back to central supply for sterilization.

Cycle of Infection

There are four main factors that contribute to the spread of disease, and they are commonly referred to as the cycle of infection:

1. Pathogenic organism
2. Reservoir of infection
3. Means of transmission
4. Susceptible host

Susceptible host　　　　Pathogenic organism

Means of transmission　　　　Reservoir of infection

Routes of Transmission

1. **Direct contact transmission:** Host is infected person placing organisms in direct contact with susceptible tissue.

   - **Indirect contact:** An inanimate object containing pathogenic organisms is placed in contact with a susceptible person.

2. **Airborne transmission:** Host inhales organisms, such as active tuberculosis, in respiratory droplets or dust.

3. **Droplet transmission:** This includes coughs, sneezes, or other methods of spraying onto a host.

4. **Common vehicle transmission:** Transmitted by infected items such as medications, food, water, devices, and equipment.

5. **Vector transmission:** Animals, such as mosquitoes and ticks, transmit infectious organisms.

6. **Fomites:** These are objects that come in contact with pathogens, such door handles, x-ray tables, Bucky diaphragms, etc.

Standard Precautions

- Always wear gloves.
- Protect clothing by wearing a protective gown or plastic apron if there is any chance of coming into contact with bodily substances.
- Masks and/or eye protection must be worn if a chance of splashing exists.
- Wash hands for a minimum of 15 seconds; this is the most effective method to prevent the spread of infection.
- Needles should never be recapped.
- Contaminated items must be disposed of properly.
- Uncapped syringe units and all sharps must be discarded in biohazard containers.

- Use appropriate protective gear when performing CPR.

Types of Isolation Precautions

Transmission-Based Precautions

- **Airborne precautions:** Masks and gowns are required, as well as negative pressure ventilation in the room.
- **Droplet precautions:** Particulate masks are required.
- **Contact precautions:** Masks, gloves, and gowns are required.

Working with Isolation Patients

- Two technologists should be present to avoid contamination of equipment.
- Clean with sanitary disinfectant immediately after exam.
- Disposal of all items in appropriate containers is a must.

Patient Transfer

Always remember to check identification bracelets to ensure that the correct patient is being transferred. Ask the patient to tell you her name and date of birth and cross-reference this with your information. Let your patient know that you are going to be moving him to the stretcher and that he may experience some discomfort during the move. Proper body mechanics must always be used during all patient transfers.

Body Mechanics

- Use a broad and stable base of support (feet apart with one foot in front of the other).
- Keep the load balanced and close to your body.
- Keep your back straight, avoid twisting, and work at a comfortable height; when reaching near the floor, bend your knees.
- Use thigh and abdominal muscles when possible (better to push than pull or lift).

Transfer from Wheelchair to X-Ray Table

- Wheelchair should be parallel to table.
- Apply brakes.
- Place step stool nearby.
- Assist the patient to a standing position.
- With patient's back against table, place patient in sitting position on the edge of the table.

Transfer from Table to Wheelchair

- Apply brakes.
- Assist patient to a sitting position on the table.
- Pivot legs so they are on the edge of the table.
- Stand facing the patient and assist patient onto chair with your knees bent.
- Position foot and leg rests into place.
- Cover patient's lap with a sheet.

Transfer from Stretcher to Table

- Place stretcher parallel to the x-ray table and lock wheels.
- Do not attempt to transfer the patient from the stretcher to x-ray table without the help of a coworker.
- One person assists in supporting the neck and shoulders, and one person supports the knees and pelvis.
- Transfer sheet and slide board should be used when available.

Patient Assessment and Comfort

Evaluate the condition of your patient and adjust pillows or sponges to allow for a smooth examination with minimal discomfort. Be sure to evaluate the elasticity of your patient's skin, as plastic wrapped sponges may irritate or rip sensitive skin.

Patients with compromised motor control may need help undressing and dressing. Clothing should be removed from the uninjured side first.

Recognizing Specific Conditions

- Dyspnea occurs when there is difficulty breathing.
- Quadriplegia is when all limbs are paralyzed; caused by cervical trauma.
- Paraplegia is when a pair of limbs is paralyzed; caused by lower cord damage.
- Hemiplegia is when one side of the body is paralyzed; caused by stroke (CVA) or illness.
- In emergency situations with cervical or other spinal injuries, patients may be moved and stabilization devices removed only in the presence of a physician.
- Fractured long bones should be supported both proximally and distally when moving patients.

Types of Fractures

- **Compression:** Fracture that causes compacting of the bone, decreasing the length and width.
- **Impacted:** When one fragment is firmly driven into the other.
- **Overriding:** The slipping of either part of a fractured bone past the other.
- **Comminuted:** When bone is splintered or crushed.
- **Incomplete:** When fracture does not entirely destroy the continuity of the bone.
- **Double:** The fracture of a bone in two places
- **Indirect:** When a fracture is away from the site of injury.
- **Compound:** A fracture in which the bone is sticking through the skin.
- **Linear:** A fracture that runs parallel to the long axis of a bone.
- **Depressed fractures:** Fracture of the skull in which a fragment is depressed.
- **Greenstick:** Bending of bone resulting in a partial fracture; usually in children.
- **Hang mans:** Fracture through the pedicles of the axis (C-2) with or without subluxation of the second or third cervical vertebra.

- **Colles:** Fracture of the distal radius with posterior displacement.
- **Boxer's:** Fracture of metacarpal neck.
- **Torus:** Impacted fracture with bulging of the periosteum.
- **Subluxation:** Partial dislocation.

Patient Monitoring and Support Equipment

- Nasogastric tubes are used to feed or suction the stomach. Clamps or plastic plugs at the end of the tube should be in place. Universal precautions should be adhered to when cleaning, opening, or closing NG tubes.
- Nasoenteric tubes are introduced further into the intestinal tract, usually to remove fluid/gas after surgery. Same precautions as nasogastric tubes should be taken into consideration.
- Urinary catheters function to drain and collect urine. Remember to clamp off the catheter before genitourinary studies and use sterile technique. Bladder infections are common with catheters. (Don't forget to unclamp after the study is over.)
- Swan-Ganz, Hickman catheters, and CVP lines are used to monitor pressures in the cardiovascular and pulmonary systems.
- Chest tubes are used to remove fluid or air from the pleural space. Chest tubes should be handled with extreme care as an involuntary disconnection involves a risk of atelectasis of lungs.
- Traction is used for orthopedic patients to promote proper bone alignment and immobilize the part for healing. Weights should not be moved or dislodged.
- A tracheotomy is a sterile, surgical opening made in the trachea when an alternate airway is necessary. When a tube is inserted and placed through the opening, it is then called a tracheostomy.
- IV lines and pumps are not to be tampered with. Radiographers are not trained in the flow rate of IV pharmaceuticals.

- Oxygen supplementation comes in many forms (mask, nasal cannula, or tent). Oxygen is considered to be a medication and must be ordered by the patient's physician. The flow is measured in L/min. Normal flow of oxygen is 3–5 L/min. Nasal cannula delivers 20–60% oxygen, while a re-breather mask (mask with airbag attached) delivers 60–90%. Tents are commonly used for pediatric patients. Masks generally deliver higher flow rates.
- Suction is particularly important for patients with low levels of consciousness. Clearance of the airway is vital. Remember that a compromised patient needs to be rolled on his or her side to prevent aspiration of vomit, and universal body fluid precautions should be utilized.
- Wheelchairs should be locked to maintain patient safety.
- Side rails should be up whenever a patient is on the gurney.
- A ventilator is a mechanical breathing device to assist the patient when she cannot breathe on her own.

Vital Signs

Temperature

The patient's temperature should be measured by utilizing a thermometer, and an accurate reading can be obtained in any of three locations:

1. Orally (mouth, under tongue): 98.6°F
2. Axillary (armpit): 97.6°F
3. Rectally (rectum): 99.6°F

Elevated or decreased temperatures are listed below and termed as such:

- Hypothermia is below normal temperature.
- Pyrexia is a fever greater than 98.6° and less than 105.8°F.
- Hyperpyrexia is a life-threatening temperature, above 105.8°F.

Pulse

A pulse greater than its maximum limit is considered tachycardic. A pulse less than its minimum limit is considered brachycardic.

Normal heart rates are as follows:

- infants (birth to 1 years old): up to 120 bpm
- children (ages 1–10 years old): 90–100 bpm
- adults: 60–100 bpm

Where to Palpate

To obtain a pulse, one of the following locations should be used:

- temporal artery (lateral skull)
- carotid artery (neck)
- radial artery (wrist)
- dorsalis pedia (foot)
- popliteal artery (knee)
- brachial artery (elbow)
- femoral artery (groin)

Respiration

The normal rate of respiration is 12 to 16 breaths per minute.

Blood Pressure

Blood pressure is categorized as systolic or diastolic. This is measured with a sphygmomanometer on the inside of the elbow at the brachial artery, which is the upper arm's major blood vessel that carries blood away from the heart.

- Systolic pressure is measurement of the contraction of the heart.
- Diastolic pressure is measurement of the relaxation of the heart.
- Diastolic pressure greater than 90 mm Hg indicates hypertension.
- Diastolic pressure less than 50 mm Hg indicates hypotension.

Medical Emergencies

Fainting or syncope: Assist patient to safe surface while supporting skull.

Hyperglycemia: High blood sugar (diabetes); breath smells fruity, skin is flushed, and patient is usually very thirsty. It has a slow onset.

Hypoglycemia: Low blood sugar, sweaty, clammy, cold, nervous, confused, and possible blurred vision. It has a rapid onset.

Seizure: Seizures are symptoms of a brain problem. They happen because of sudden, abnormal electrical activity in the brain. Do not attempt to control seizure; allow seizure to take its course and support patient's head. Below are two types of seizures:

1. **Petit mal:** A petit mal seizure is the term commonly given to a staring spell, most commonly called an absence seizure. It is a brief (usually less than 15 seconds) disturbance of brain function due to abnormal electrical activity in the brain.
2. **Grand mal:** Also known as a tonic-clonic seizure, it features a loss of consciousness along with violent muscle contractions. It is the type of seizure most people picture when they think about seizures in general. There are often loud cries from the patient and contraction of abdomen and chest. Remove hazardous objects nearby and move patient to floor with padding for head, if possible.

Arterial bleeding: Pulsing blood from wound site; use direct compression on site or proximal compression.

Venous bleeding: Less forceful pulsing; compress on site.

Orthostatic hypotension: This is caused by a rapid reduction in blood pressure (sitting up suddenly from lying down); patient feels faint or light-headed.

Cardiac arrest: This is the cessation of heart function. Code team must be called and CPR must begin immediately. Radiographer must obtain code cart.

Respiratory arrest: This is the cessation of breathing.

Shock (life-threatening): This involves circulatory failure; not enough blood pressure to supply vital tissues with proper amounts of oxygen. Types of shock include:

- **hypovolemic:** loss of blood due to trauma, internal, or external hemorrhage
- **septic:** severe infection
- **neurogenic:** caused by head or spinal trauma
- **cardiogenic:** caused by heart failure
- **anaphylactic:** caused by an injection with foreign proteins such as bee stings, contrast agents, etc.

Trauma: Serious and potentially life-threatening injuries. At no time should a trauma patient be left alone in an examination room. Trauma patients must be handled with the most delicate care, as any wrong movement may add harm to the patient's injury. Neck brace and back boards shall not be removed until clearance is given by the attending physician post x-rays. All trauma patients should be classified in one of four areas of consciousness:

1. Alert and oriented
2. Dizzy or drowsy
3. Unconscious
4. Comatose

Cardiopulmonary Resuscitation (CPR)

Cardiopulmonary resuscitation (CPR) is a lifesaving technique useful in many emergencies, including cardiac arrest or near drowning, in which someone's breathing or heartbeat has stopped. There are two elements that make up the CPR technique: chest compressions combined with mouth-to-mouth rescue breathing.

Remember the ABCs

1. **Airway:** Clear the airway.
2. **Breathing:** Breathe for the person.
3. **Circulation:** Restore blood circulation with chest compressions.

Adults:

1. You should compress the chest 1.5 to 2 inches in depth.
2. There should be 100 beats per minute.
3. There should be 30 chest compressions to every two breaths.

Children:

1. You should compress the chest 1 to 1.5 inches in depth.
2. There should be 100 beats per minute.
3. There should be 30 chest compressions to every two breaths, using one rescuer.
4. There should be 15 chest compressions to every two breaths, using two rescuers.

Infants:

1. Place two or three fingers on the center of the chest, just below the nipple line.
2. Compress the chest about a 0.5 to 1 inch.
3. Use smaller breaths for infants.
4. There should be 30 chest compressions to every two breaths, using one rescuer.
5. There should be 15 chest compressions to every two breaths, using two rescuers.
6. Goal is 100 beats per minute.

Fire

In the case of a fire, you should always remember the acronym RACE:

1. Rescue
2. Alarm
3. Confine
4. Extinguish

Patient Preparation for Examination

Gastrointestinal or Urinary System Examination

- Patient should be on a low-residue diet.
- NPO is indicated for 8 to 12 hours prior to examination.
- Cleansing enemas are used to clean the GI system.

Contrast Media

Contrast agents are used to enhance radiographic contrast or differentiation of soft tissues by altering the attenuation of the x-ray beam. Kidney function testing must be performed prior to any IV contrast administration.

Negative Contrast Agent

Air is the most common use of negative contrast. Negative contrast results in a much higher density on the radiograph and is combined with a positive agent on double contrast studies such as barium enemas. A negative contrast reduces the attenuation to increase the radiographic density.

Positive Contrast Agents

Positive contrast agents, such as iodine and barium with an atomic number of 53, increase the attenuation and decreases the radiographic density. This provides an increase in contrast between the structure in focus and its surroundings.

Barium

Administered in the form of a barium sulfate, this is most commonly used for both an upper GI series and esophograms and is not absorbed by the body. Barium should be served cold enough for the patient to tolerate. Room temperature barium sulfate is just not palpable. When preparing a powderlike mixture for barium enema studies, warm water at a temperature of no higher than 100°F must be used.

Aqueous Iodine Compounds

Aqueous iodine compounds are most commonly used for contrast studies of the gastrointestinal tract and when the possibility of a ruptured appendix or perforated ulcer is present.

Iodinated Contrast Media

Ionic contrast agents provide for a higher risk of contrast reactions. Ionic contrasts are composed of both positively and negatively charged ions.

Nonionic Contrast Agents

Nonionic contrasts do not ionize into separate positive and negative charges. This is their main advantage over ionic contrasts, and they have a much lower risk of contrast reactions.

Contraindications

The best predictor of reaction is a thorough patient history and previous allergic reactions the patient might have experienced. Be sure to always check the patient's chart and ask if the patient has ever had a reaction to contrast media before the procedure begins. If he or she responds yes, the primary care physician must be contacted to revise the order if needed. Less than 5% of patients experience a reaction (variable), and of these only 5% experience a severe or life-threatening reaction.

Adverse Reactions

Local: Well visualized on the surface

Extravasation: Contrast leaks into interstitial tissues at injection site, causing pain, redness, and swelling. Too much contrast extravasated under the tissue may require surgical intervention to repair. Always apply a warm compress to the area and elevate extremity.

Systemic: Not well visualized; internal sensations will appear to the patient.

Symptoms of Adverse Reactions

- restlessness and apprehension
- increased heart rate
- pallor with weakness and/or altered mental status
- cool, clammy skin
- decreased blood pressure, shortness of breath, itchy

Classifications of Adverse Reactions

- Mild reaction: Roll patient on side, provide emesis basin, and observe.
- Moderate reaction: Administer antihistamine.
- Severe reaction: Administer antihistamine and call cardiac code.
- Always document the patient's reactions in his or her chart and complete an incident report in the event a reaction should occur.

Patient History

The patient's history essentially begins when the radiographer introduces himself or herself and verifies the patient's name and date of birth. The patient's history is only as good as the radiographer's questions. The answers provide the radiographer with important information about the extent of the patient's injury, and this will assist in the radiologist's interpretation of the films. Below is a list of questions that may assist you when interviewing your patients:

1. Why did your doctor order this exam?
2. How and when did you get hurt?
3. Do you have any pain or numbness? Is so, where?
4. Do you a fever? If so, how long and at what temperature?
5. Any shortness of breath, coughing, or history of COPD?
6. Any nausea, vomiting, or diarrhea?
7. Any prior surgeries?
8. Is there any chance you are pregnant?
9. Date of last menstrual period?

Assessing the Changing Conditions of a Patient

Visually observe the patient, look to see if the patient has a change in skin and lip tone. Continuously ask questions such as, How do you feel? Are you dizzy or uncomfortable? If the patient feels cold to the touch, offer a blanket. The patient's blood pressure may be low, causing a lowering of body temperature. Small suggestions such as these will certainly help when assessing your patient.

Sequencing and Scheduling Radiographic Procedures

Sequencing: If a patient is scheduled to have more than one study, think about what would least impact the next study. Consider the following:

- exams not requiring contrast media
- thyroid studies (iodine uptake)
- urinary tract exams
- biliary exams
- gastrointestinal exams

Scheduling: Which patients should have priority?

- surgical
- diabetic (NPO)
- emergency

- pediatric
- geriatric

Patient Education

There are several factors that come into play when we speak about patient education. First and foremost, as a professional, it is among your duties and responsibilities to make sure your patients are well informed and made comfortable so their experience is a pleasant one. Below is a list that should be adhered to when educating a patient:

- If you have literature about the exam, have them read it while they are in the waiting room. This usually answers a lot of their questions.
- Give a full detailed explanation of the examination.
- Verify consent.
- Explain how the procedure will be performed.
- Explain why the exam is being performed and what you are looking for.
- Give an estimate of how long the test should take.
- Tell them when they can anticipate results to be ready.

Post-Examination Instructions

Always inform your patients that if they experience any post-procedural complications, they should immediately contact their primary care physician. They should also increase their fluid intake (eight glasses of water), as this will dilute contrast media and increase output much quicker.

Routes of Administration

There are many ways to administer contrast media, based on the procedure and patient's diagnosis. It is always important to verify that you have the right patient, right drug, right amount, right route, and right time. The minimum height of an IV or barium bag is always 18 inches above the insertion/injection site. The means of administration are as follows:

- IV (IVP)
- rectally; patient in Sims position (barium enema)
- orally (UGI, small bowel series, swallowing studies)
- direct injection (surgical cholangiograms)

Venipuncture

Venipuncture needles are best described by the size of the gauge. The higher the gauge, the smaller the diameter of the opening, and vice versa. The most common sites of injection are the median cephalic and basilic veins. The following steps must be performed to obtain a successful venipuncture procedure:

1. Wash hands prior to injection.
2. Always wear gloves.
3. Place tourniquet above site of injection.
4. Select and palpate vein.
5. Cleanse site of injection with alcohol swab.
6. Insert needle into vein with bevel toward heart.
7. Confirm return of blood prior to injection.
8. Begin hand injection with 10 cc of contrast, and observe site of injection. If site is free of extravasation and patient can tolerate contrast, continue injection.
9. Remove tourniquet.
10. Upon completion of injection, remove and discard needle in appropriate sharps container.
11. Place gauze on site of injection and compress site for one minute.
12. Place bandage on site.
13. Wash hands.

Chapter 9 Review Questions

1. The unlawful touching of an individual without his or her consent is considered
   a. battery.
   b. assault.
   c. invasion of privacy.
   d. tort.

2. When correctly identifying a patient, which method(s) of verification should be used?
   1. Ask the patient his/her name.
   2. Check patient's wristband for accuracy.
   3. Check chart only.
   4. Find out date of birth.

   a. 1 only
   b. 1 and 2 only
   c. 3 only
   d. 1, 2, and 4 only

3. Which factors must be present to contribute to the spread of a disease?
   1. reservoir of infection
   2. pathogenic organism
   3. means of transmission
   4. susceptible host

   a. 1 and 2 only
   b. 1 and 4 only
   c. 3 and 4 only
   d. all of the above

4. Prior to the beginning and upon the end of a procedure, how long should you wash your hands to be free of germs?
   a. 15 seconds
   b. 25 seconds
   c. 45 seconds
   d. 60 seconds

5. If a patient presents to the hospital with a life-threatening temperature above 105.8°F, what is this condition called?
   a. hypersthenic
   b. hyperglycemic
   c. hyperpyrexia
   d. hyperthermia

6. If your only method of sterilization is boiling, how long should you boil an item to make it usable in a sterile environment?
   a. 12 minutes
   b. 15 minutes
   c. 20 minutes
   d. 22 minutes

7. If a patient is placed on airborne isolation, which of the following protective garment(s) must be worn prior to entering the room?
   1. mask
   2. gloves
   3. booties
   4. gown

   a. 1 and 2 only
   b. 1 and 4 only
   c. 1, 2, and 4 only
   d. all of the above

8. The term *syncope* refers to
   a. hyperventilation.
   b. diabetic shock.
   c. seizures.
   d. fainting.

9. When performing a two-person resuscitation on a 12-year-old patient, the ratio of compressions to breaths is
   a. 30 compressions to three breaths.
   b. 15 compressions to two breaths.
   c. 20 compressions to two breaths.
   d. 30 compressions to two breaths.

10. Barium sulfate (BaSO4) is most commonly used for which of the following procedures?

    1. upper GI series
    2. IVP
    3. esophograms
    4. ERCP

  a. 1 only
  b. 1, 3, and 4 only
  c. 1 and 3 only
  d. all of the above

11. Iodine has an atomic number of
  a. 53.
  b. 63.
  c. 52.
  d. 48.

12. Prior to a small bowel series, the patient must be NPO for how many hours?
  a. 6–12
  b. 8–10
  c. 8–12
  d. 6–8

13. When working with isolated patients, how many technologists should be present to avoid contamination of equipment?
  a. one
  b. two
  c. three
  d. four

14. When transferring a patient from an exam table to a wheelchair, in which direction should the wheelchair face in conjunction with the table?
  a. perpendicular
  b. horizontal
  c. vertical
  d. parallel

15. The term *stroke* is commonly referred to as
  a. CVA.
  b. DVT.
  c. TIA.
  d. ARC.

16. A pulse greater than 100 beats per minute is considered
  a. brachycardic.
  b. tachycardic.
  c. hypoglycemic.
  d. hypertensive.

17. When measuring blood pressure, which artery should be used to obtain the best reading?
  a. femoral
  b. dorsal
  c. brachial
  d. popliteal

18. Parental informed consent is required on all patients under the age of
  a. 16.
  b. 17.
  c. 18.
  d. 21.

19. A patient's normal rate of respiration should be
  a. 12–16 breaths per minute.
  b. 16–20 breaths per minute.
  c. 15–19 breaths per minute.
  d. 8–10 breaths per minute.

20. Nasoenteric tubes are used for which of the following?
  a. to feed patient
  b. to remove fluid or gas after surgery
  c. to administer medications
  d. to drain air or fluid from pleural space

21. Which of the following arteries may be used to obtain a patient's pulse?

    1. carotid
    2. brachial
    3. dorsalis pedia
    4. femoral

a. 2 only
b. 2 and 3 only
c. 1, 4, and 2 only
d. all of the above

22. Which of the following is true when referring to positive contrast agents?
a. It decreases the attenuation and decreases the density.
b. It increases the attenuation and decreases the density.
c. It increases the attenuation and increases the density.
d. It decreases the attenuation and increases the density.

23. The term *diastolic* means
a. measurement of the heart during contraction.
b. fever.
c. heart failure.
d. measurement of the heart during relaxation.

24. If a patient is unable to chew or cannot tolerate solid foods, which of the following is used to feed the patient?
a. nasoenteric tube
b. Swan Ganze catheter
c. nasogastric tube
d. central line

25. When performing a procedure in a sterile environment, the operator and surgeon should only pass each other in which direction?
a. back to back
b. front to back
c. side to side
d. back to front

26. The term *dyspnea* means
a. difficulty sleeping.
b. difficulty eating.
c. difficulty breathing.
d. difficulty seeing.

27. A sphygmomanometer is used to measure
a. temperature.
b. infection.
c. pulse.
d. blood pressure.

28. Nasocomial infections are acquired in which type of environment?
a. school
b. hospital
c. doctor's office
d. park

29. If you witness a patient going into a seizure for 15 seconds or less, what kind of seizure is this considered to be?
a. grand mal
b. petit mal
c. comatose
d. neurogenic shock

30. If a patient presents with low blood sugar and a rapid onset, he or she is considered to be
a. hyperglycemic.
b. hyposystemic.
c. hypoglycemic.
d. tachycardic.

31. The injurious touching of an individual without his or her consent is considered
a. battery.
b. tort.
c. slander.
d. assault.

32. In the event that a patient becomes incapacitated, which document must have been completed to assure proper and adequate care is provided?
a. advanced directive
b. informed consent
c. verbal consent
d. power of attorney

33. The most common fracture of the wrist is called
a. a compound fracture.
b. an indirect fracture.
c. a greenstick fracture.
d. a boxer's fracture.

34. Which route of transmission would best describe a disease or infection which is transmitted via an animal?
a. airborne
b. vector
c. droplet
d. indirect

35. Which of the following is NOT considered nonverbal communication?
a. facial expressions
b. body language
c. speaking clearly
d. eye contact

36. An advanced directive should be filled out by the patient before he or she
a. becomes incapacitated.
b. is discharged from the hospital.
c. returns to work.
d. is admitted.

37. If a 48-year-old patient has a mammogram, how long should the films be stored?
a. 5 years
b. 5–7 years
c. 18– 21 years
d. life span of patient

38. Which of the following items must be present on a radiograph?
1. name
2. date
3. left or right markers
4. medical record number

a. 1 and 2 only
b. 1, 2, and 3 only
c. 1, 2, and 4 only
d. all of the above

39. What does the term *res ipsa loquitur* mean?
a. Let the master answer.
b. to digress
c. The thing speaks for itself.
d. The master speaks for others.

40. The normal pulse rate of an adult should be between
a. 60–100 breaths per minute.
b. 50–90 beats per minute.
c. 40–80 breaths per minute.
d. 60–100 beats per minute.

41. A sterile surgical opening in the trachea is called a
  a. tracheostomy.
  b. tracheotomy.
  c. tracheocarcinoma.
  d. tracheostoma.

42. An extravasation is considered to be what kind of adverse reaction?
  a. local
  b. cardiogenic
  c. systemic
  d. anaphylactic

43. Which type of contrast separates into positive and negative charges?
  a. aqueous
  b. ionic
  c. nonionic
  d. barium

44. The minimum height of an IV bag should be no less than
  a. 24 inches.
  b. 18 inches.
  c. 12 inches.
  d. 10 inches.

45. What temperature should the water be when mixed with powder for a barium enema?
  a. 98.6°F
  b. 99°F
  c. 100°F
  d. 100.4°F

46. A trauma patient should only be left in the room alone in which of the following cases?
  a. when accompanied by a family member
  b. when he or she says he or she will be fine to be left alone
  c. when the pain is gone
  d. never

47. A patient with a diastolic reading of less than 50 mm Hg is considered to be
  a. hypertensive.
  b. hypotensive.
  c. hyperoxic.
  d. orthostatic hypotensive.

48. The best method for preventing aspiration is to
  a. lay patient supine.
  b. place the patient in a Trendelenberg position.
  c. roll patient onto his or her side.
  d. none of the above.

49. When performing a barium enema, the patient must be placed in which position?
  a. SIMS
  b. recumbent
  c. oblique
  d. prone

50. What is the main purpose of a Swan Ganz catheter?
  a. to inject contrast
  b. to administer medications
  c. to monitor pressure
  d. to remove fluid from pleural cavity

51. What does the term *systolic* mean?
    a. measurement of heart during relaxation
    b. measurement of oxygen per minute
    c. measurement of the lungs during respiration
    d. measurement of the heart during contraction

52. Which device is utilized to promote proper bone alignment and immobilization as part of proper healing?
    a. retractor
    b. splint
    c. traction
    d. cast

53. When lifting a heavy object, it is always important to do which of the following?
    1. Bend your knees.
    2. Twist your back.
    3. Ask for assistance.
    4. Do it yourself.

    a. 4 only
    b. 1 and 3 only
    c. 1, 2, and 3 only
    d. all of the above

54. Quadriplegia is caused by blunt trauma to the
    a. lower cord.
    b. cervical spine.
    c. thoracic spine.
    d. lumbar vertebra.

55. Which of the following routes of transmission would cause an inanimate object containing pathogenic organisms to be placed in contact with a susceptible person?
    a. direct contact
    b. airborne
    c. indirect contact
    d. formites

56. A sterile item must always be sent to central supply for sterilization unless the item is which of the following?
    1. opened but never used
    2. partially used
    3. contaminated by a nonsterile person

    a. 1 only
    b. 2 and 3 only
    c. 1 and 2 only
    d. none of the above

57. When using dry heat as a source of sterilization, what temperature and duration of time must be used to completely remove all microorganisms?
    a. one to three hours, 300°F+
    b. one to six hours, 300°C+
    c. one to five hours, 300°F+
    d. one to six hours, 300°F+

58. A Do Not Resuscitate order is generated to prevent which of the following actions from taking place?
    a. code call
    b. rapid response
    c. consultation
    d. clergy call

59. If a patient loses a limb due to the conscious disregard or selfish actions of an individual, this is considered
    a. gross violence.
    b. gross evidence.
    c. gross negligence.
    d. contributory negligence.

60. The type of tonic-clonic seizure which features a loss of consciousness and violent muscle contractions is also known as
   a. petit mal.
   b. grand mal.
   c. cum mal.
   d. lars mal.

61. In which of the correct orders should the following patients be scheduled based on their priority?
   a. surgical, pediatric, diabetic (NPO), geriatric, emergency
   b. pediatric, emergency, diabetic (NPO), geriatric, surgical
   c. emergency, diabetic (NPO), surgical, pediatric, geriatric
   d. surgical, diabetic (NPO), emergency, pediatric, geriatric

62. When performing a surgical cholangiogram, which of the following routes of administration should be utilized?
   a. direct injection
   b. intravenous injection
   c. intramuscular injection
   d. subcutaneous injection

63. Which is the most commonly used artery when performing a venipuncture procedure?

   1. cephalic
   2. basilic
   3. femoral
   4. popliteal

   a. 1 and 2 only
   b. 2 only
   c. 1 and 3 only
   d. 1, 2, and 4 only

64. The normal blood pressure range for an adult is
   a. 100/80 mm Hg.
   b. 120/60 mm Hg.
   c. 80/120 mm Hg.
   d. 120/80 mm Hg.

65. In the event that a patient experiences a severe loss of blood, there is a great possibility that he or she may go into which type of shock?
   a. hyperventilation
   b. hypovolemic
   c. hypodermic
   d. hypergenic

66. The use of restraints without the proper consent and authorization is called
   a. battery.
   b. assault.
   c. false imprisonment.
   d. free will.

67. The Latin term *respondeat superior* means which of the following?
   a. Let the king answer.
   b. The thing speaks for itself.
   c. Let the master answer.
   d. Let us answer for the master.

68. How long should radiographs be stored for a 22-year-old patient?
   a. for the life of the patient
   b. one year
   c. three years
   d. seven years

69. If a patient presents in the emergency room with multiple bee stings and seems to be going into shock, what type of shock would best describe this situation?
- **a.** cardiogenic
- **b.** neurogenic
- **c.** septic
- **d.** anaphylactic

70. The term *hyperglycemia* refers to which of the following?
- **a.** high blood sugar
- **b.** high urine output
- **c.** high creatinine levels
- **d.** high blood pressure

71. When scheduling patients for their procedures, at what time of day would you schedule a barium enema?
- **a.** before noncontrast studies are performed
- **b.** mixed in with noncontrast studies
- **c.** after noncontrast studies are performed
- **d.** the next day

72. When placing a patient on or off oxygen, what would be the normal flow rate to use?
- **a.** 2–4 L per minute
- **b.** 3–5 L per minute
- **c.** 1–3 L per hour
- **d.** 4–6 L per second

73. What is the main purpose of a chest tube?
- **a.** to retain fluid or air in the pleural space
- **b.** to supply a central line for IV injections
- **c.** to remove air or fluid from pleural space
- **d.** to feed patient

74. Air is considered which type of contrast agent?
- **a.** positive
- **b.** negative
- **c.** iodinated ionic
- **d.** iodinated nonionic

75. The acronym HIPAA means
- **a.** Health Insurance Plan and Accountability Act.
- **b.** Health Information Privacy and Accountability Act.
- **c.** Health Information Portability and Accountability Act.
- **d.** Health Insurance Portability and Accountability Act.

76. If a patient is in cardiac arrest, the heart is
- **a.** in relaxation.
- **b.** in cessation.
- **c.** in contraction.
- **d.** pumping.

77. Surgical asepsis is considered to be which of the following?
- **a.** the spreading of organisms during surgery
- **b.** the complete removal of organisms from equipment and environment
- **c.** the surgical removal of microorganisms
- **d.** the surgical removal of organisms from patient

78. A written document that results in a defamation of character or loss of reputation is considered
- **a.** slander.
- **b.** tort.
- **c.** libel.
- **d.** invasion of privacy.

79. Who was responsible for passing the Health Insurance Portability and Accountability Act in 1996?
- **a.** U.S. Congress
- **b.** House of Representatives
- **c.** secretary of state
- **d.** Department of Homeland Security

80. Which of the following is true regarding negative contrast agents?
- **a.** They decrease the attenuation and decrease the density.
- **b.** They increase the attenuation and decrease the density.
- **c.** They increase the attenuation and increase the density.
- **d.** They decrease the attenuation and increase the density.

81. When the possibility of a perforated ulcer or ruptured appendix exists, which type of contrast should be used?
- **a.** ionic contrast
- **b.** nonionic contrast
- **c.** aqueous iodine compounds
- **d.** barium sulfate

82. Why is it so important for your patient to drink a minimum of eight cups of water after any IV procedure?
- **1.** It increases urine output.
- **2.** It dilutes contrast.
- **3.** It produces insulin.
- **4.** It increases kidney function.

- **a.** 1 and 2 only
- **b.** 1, 3, and 4 only
- **c.** 1, 2, and 4 only
- **d.** all of the above

83. Which statement best describes the diameter of the opening of a needle?
- **a.** The lager the gauge, the smaller the diameter of the opening.
- **b.** The smaller the syringe, the larger the diameter of the opening.
- **c.** The smaller the gauge, the larger the diameter of the opening.
- **d.** The color-coded gauges have the largest diameter opening.

84. Which type of contrast media provides for a higher risk of reactions?
- **a.** ionic
- **b.** nonionic
- **c.** gadolinium
- **d.** aqueous iodine

85. What does the acronym RACE stand for?
- **a.** rescue, assist, call, exit
- **b.** rescue, alarm, confine, extinguish
- **c.** ring, assist, close doors, exit
- **d.** rescue, alarm, confine, evacuate

86. In case of written error during documentation, which method should be used to void the error?
- **a.** Erase the error.
- **b.** White out the error.
- **c.** Draw a single line through the error.
- **d.** Write over the error and initial.

87. A major advantage of barium sulfate is that
- **a.** it is absorbed well in the body.
- **b.** it does not get absorbed by the body.
- **c.** it contains essential nutrients.
- **d.** it appears radiolucent on a radiograph.

88. Common vehicle transmission is caused by infected items such as
    a. medications, food, water, and equipment.
    b. sneezing and coughing.
    c. dust.
    d. ticks and mosquitoes.

89. Who is legally allowed to order an examination on a patient?
    a. nurse practitioner
    b. medical doctor
    c. technologist
    d. medical assistant

90. Nausea, vomiting, hives, and flushed skin are all symptoms of which type of reaction?
    a. anaphylactic
    b. cardiogenic
    c. neurogenic
    d. hyperglycemic

91. How deep should the chest be compressed during cardiopulmonary resuscitation of a child?
    a. 0.5–1.0 inches
    b. 1.0–1.5 inches
    c. 1.5–2.0 inches
    d. 2.0–2.5 inches

92. Where is a dorsalis pedia pulse taken?
    a. groin
    b. wrist
    c. foot
    d. knee

93. Which of the following statements is NOT true regarding the discarding of needles?
    a. Used needles should always be considered contaminated.
    b. All unused needles should always be re-capped.
    c. IV catheters must be discarded in sharps container.
    d. Never break the needle off into a sharps container.

94. A consistent body temperature reading of 97.0° is considered to be
    a. hyperthermia.
    b. pyrexia.
    c. hypothermia.
    d. hyperpyrexia.

95. A mechanical breathing device that assists patients when they cannot breathe on their own is called
    a. tracheostomy.
    b. re-breather mask.
    c. oxygen tank.
    d. ventilator.

96. A fracture through the pedicles of the axis C-2 is called
    a. Colles.
    b. hangman's.
    c. compound.
    d. depressed.

97. In the event a trauma patient arrives in the emergency room secured in a neck brace and backboard, when is it safe to remove both items?

a. immediately after radiographs are completed

b. immediately after patient requests it off

c. upon review of films and cleared by EMT

d. upon review of films and cleared by the attending physician

98. A normal pulse range for children one to ten years of age should be

a. 80–110.

b. 90–100.

c. 100–110.

d. 70–80.

99. Which type of shock may a patient experience if he or she acquires a severe infection?

a. septic

b. anaphylactic

c. cariogenic

d. neurogenic

100. Which definition best describes an impacted fracture?

a. The bone is splintered or crushed.

b. The bone is fractured in two places.

c. One fragment is firmly driven into the other.

d. The distal radius is fractured with posterior displacement.

Chapter 9 Answers

1. **a.** Unlawfully touching an individual without his or her consent is battery and is punishable in a court of law.

2. **d.** When identifying a patient, it is always critical to verify by having the patient tell you his or her name and date of birth as well as checking his or her wristband.

3. **d.** All of the above must be present to form the cycle of disease.

4. **a.** You must wash your hands for at least 15 seconds to be germ-free.

5. **c.** Hyperpyrexia is the life-threatening temperature above 105.8°F.

6. **a.** When boiling an item for sterilization, each item must be boiled for a minimum of 12 minutes.

7. **c.** Airborne isolation requires all technologists entering the patient's room to wear a mask and gown for protective purposes. Gloves are a universal precaution and are required for all examinations.

8. **d.** Fainting is the most common term for syncope.

9. **b.** When performing two-person CPR, the ratio is 15:2.

10. **c.** Upper gastrointestinal series and esophograms use barium sulfate.

11. **a.** Both iodine and barium have an atomic number of 53.

12. **b.** The patient must be NPO for 8–10 hours prior to the procedure.

13. **b.** Two people should always work with isolated patients, one clean and one dirty, to avoid contamination of equipment.

14. **d.** The wheelchair must always be parallel.

15. **a.** CVA is also known as a stroke.

16. **b.** Tachycardic is a pulse of over 100 bpm.

17. **c.** The brachial artery should be used to obtain the most accurate reading.

18. **c.** Any person(s) under the age of 18 years old is considered a minor and cannot sign a legal document.

19. **a.** Twelve to 16 breaths per minute is the normal respiration rate.

20. **b.** Nasoenteric tubes are used to remove fluid or gas after surgery.

21. **d.** All of the above; you may obtain a pulse from any palpable artery.

22. **b.** It increases the attenuation and decreases the density.

23. **d.** Diastolic is the measurement of the heart during relaxation.

24. **c.** A nasogastric tube is used to feed or suction the stomach.

25. **a.** The operator and surgeon should only pass back to back in a sterile environment to avoid contamination.

26. **c.** Dyspnea is when a patient is having difficulty breathing.

27. **d.** A patient's blood pressure is measured by using a sphygmomanometer.

28. **b.** Nasocomial infections are only hospital acquired.

29. **b.** Petit mal is a seizure of 15 seconds or less.

30. **c.** Diabetics with low blood sugar are considered hypoglycemic.

31. **d.** Assault is the threat of injurious touching.

32. **d.** A power of attorney provides the right for another person to make decisions regarding medical care if the patient cannot communicate.

33. **d.** A boxer's fracture is the most common fracture of the wrist.

34. **b.** Vector-borne transmission occurs with animals such as mosquitoes and ticks that transmit infectious organisms to humans.

35. **c.** Speaking clearly is a form of verbal communication.

36. **a.** An advanced directive should be filled out by the patient in case he or she becomes incapacitated.

37. d. Mammograms are to be kept for the lifespan of the patient.

38. d. Name, date, medical record number, and anatomical markers should be present on a radiograph.

39. c. The Latin term means the thing speaks for itself.

40. d. The normal adult pulse rate is 60–100 bpm.

41. b. A tracheotomy is a surgical opening in the trachea to assist the patient with breathing.

42. a. Extravasation is a local adverse reaction.

43. c. Nonionic contrast separates into positive and negative charges.

44. b. An IV bag must be kept minimally 18 inches above the site.

45. c. Patients find the mixture more palatable at 100°F.

46. d. A trauma patient should never be left alone.

47. b. Hypotension is a diastolic reading of less than 50 mm Hg.

48. c. Roll patient onto his or her side to prevent aspiration.

49. a. The SIMS position is utilized during a barium enema procedure.

50. c. It is used to monitor the pressure of both the cardiovascular and pulmonary arteries.

51. d. Systolic pressure is the measurement of the heart during contraction.

52. c. Traction is used for orthopedic patients to promote proper bone alignment and immobilize the part for proper healing.

53. b. Always ask for assistance and bend your knees. Never lift with your back.

54. b. Quadriplegia is caused by blunt trauma to the cervical spine.

55. c. Indirect contact causes an inanimate object containing pathogenic organisms to be placed in contact with a susceptible person.

56. d. Open, used, and contaminated items always go back to central supply for sterilization.

57. d. One to six hours of dry heat at a temperature of 300°F+ completely removes all microorganisms.

58. a. You must respect the DNR request and never call a code.

59. c. Gross negligence is the conscious and voluntary disregard of the need to use reasonable care. It involves acts that demonstrate disregard for life or limb.

60. b. A grand mal seizure is also known as a tonic-clonic seizure. It features a loss of consciousness and violent muscle contractions.

61. d. The correct order is surgical, diabetic (NPO), emergency, pediatric, geriatric.

62. a. A direct injection is required into the duct during a surgical cholangiogram.

63. b. The basilic and cephalic arteries are most commonly used during venipuncture.

64. d. The normal blood pressure of an adult is 120/80 mm Hg.

65. b. Hypovolemic shock is when the patient experiences a severe loss of blood or plasma.

66. c. False imprisonment is the use of restraints without the proper consent and authorization.

67. c. The Latin phrase means let the master answer.

68. d. A 22-year-old patient is considered an adult; therefore, films must only be kept for seven years.

69. d. Anaphylactic shock is cause by the injection of a foreign protein such as a bee sting or contrast agent.

70. a. Hyperglycemia is a term commonly used to mean high blood sugar.

71. c. Due to the length of a barium enema study, it is always wise to schedule a case after all noncontrast studies have been performed to avoid delays.

72. b. Three to five L per minute is the normal flow rate when placing a patient on oxygen unless otherwise instructed by a physician.

73. c. The main function of a chest tube is to remove air and fluid from the pleural space.

74. a. Air is considered to be a positive contrast agent. A great example of a positive contrast is a chest radiograph.

75. d. The Health Insurance Portability and Accountability Act established legal guidelines pertaining to patient information and medical records.

76. b. Cessation means the heart stops beating.

77. b. Complete removal of organisms from equipment and environment is called surgical asepsis.

78. c. Libel is defamation of character within a written document.

79. a. The U.S. Congress is responsible for passing HIPAA.

80. d. They decrease the attenuation and increase the density.

81. c. Aqueous iodine compounds are used when there is a suspected ulcer or a ruptured appendix is present.

82. c. Water increases urine output, kidney function, and dilutes contrast.

83. a. The larger the gauge, the smaller the diameter of the opening.

84 a. Ionic contrasts provide for a higher risk of reactions.

85. b. RACE stands for rescue, alarm, confine, and extinguish.

86. c. In case of an error in documentation, draw a single line through the error and rewrite.

87. b. Barium sulfate does not get absorbed by the body.

88. a. Medications, food, water, and equipment cause common vehicle transmission.

89. b. Only an MD and PA are allowed to request/order an examination.

90. a. An anaphylactic reaction can cause nausea, vomiting, hives, flushed skin, or fever.

91. b. During CPR on a child, the chest must be compressed 1.0–1.5 inches in depth.

92. c. A dorsalis pedia pulse is taken on the foot.

93. b. Never re-cap a needle. If it was opened, discard it.

94. c. Hypothermia is when your body temperature drops below normal.

95. d. A ventilator is a mechanical breathing device that assists patients who cannot breathe on their own.

96. b. A hangman's fracture is a fracture through the pedicles of the axis C-2.

97. d. A patient is never stripped of his or her neck brace or backboard until films are reviewed and cleared by the attending physician.

98. b. 90–100 beats per minute is the normal pulse range for children one to ten years of age.

99. a. Septic shock is when a patient acquires a severe infection.

100. c. An impacted fracture is when one fragment is firmly driven into the other.

CHAPTER

10 ▶ RADIOGRAPHY PRACTICE TEST II

Radiography Practice Test II

	a	b	c	d			a	b	c	d			a	b	c	d
1.	a	b	c	d		37.	a	b	c	d		73.	a	b	c	d
2.	a	b	c	d		38.	a	b	c	d		74.	a	b	c	d
3.	a	b	c	d		39.	a	b	c	d		75.	a	b	c	d
4.	a	b	c	d		40.	a	b	c	d		76.	a	b	c	d
5.	a	b	c	d		41.	a	b	c	d		77.	a	b	c	d
6.	a	b	c	d		42.	a	b	c	d		78.	a	b	c	d
7.	a	b	c	d		43.	a	b	c	d		79.	a	b	c	d
8.	a	b	c	d		44.	a	b	c	d		80.	a	b	c	d
9.	a	b	c	d		45.	a	b	c	d		81.	a	b	c	d
10.	a	b	c	d		46.	a	b	c	d		82.	a	b	c	d
11.	a	b	c	d		47.	a	b	c	d		83.	a	b	c	d
12.	a	b	c	d		48.	a	b	c	d		84.	a	b	c	d
13.	a	b	c	d		49.	a	b	c	d		85.	a	b	c	d
14.	a	b	c	d		50.	a	b	c	d		86.	a	b	c	d
15.	a	b	c	d		51.	a	b	c	d		87.	a	b	c	d
16.	a	b	c	d		52.	a	b	c	d		88.	a	b	c	d
17.	a	b	c	d		53.	a	b	c	d		89.	a	b	c	d
18.	a	b	c	d		54.	a	b	c	d		90.	a	b	c	d
19.	a	b	c	d		55.	a	b	c	d		91.	a	b	c	d
20.	a	b	c	d		56.	a	b	c	d		92.	a	b	c	d
21.	a	b	c	d		57.	a	b	c	d		93.	a	b	c	d
22.	a	b	c	d		58.	a	b	c	d		94.	a	b	c	d
23.	a	b	c	d		59.	a	b	c	d		95.	a	b	c	d
24.	a	b	c	d		60.	a	b	c	d		96.	a	b	c	d
25.	a	b	c	d		61.	a	b	c	d		97.	a	b	c	d
26.	a	b	c	d		62.	a	b	c	d		98.	a	b	c	d
27.	a	b	c	d		63.	a	b	c	d		99.	a	b	c	d
28.	a	b	c	d		64.	a	b	c	d		100.	a	b	c	d
29.	a	b	c	d		65.	a	b	c	d						
30.	a	b	c	d		66.	a	b	c	d						
31.	a	b	c	d		67.	a	b	c	d						
32.	a	b	c	d		68.	a	b	c	d						
33.	a	b	c	d		69.	a	b	c	d						
34.	a	b	c	d		70.	a	b	c	d						
35.	a	b	c	d		71.	a	b	c	d						
36.	a	b	c	d		72.	a	b	c	d						

Radiography Practice Test II (*continued*)

101.	a	b	c	d	137.	a	b	c	d	173.	a	b	c	d
102.	a	b	c	d	138.	a	b	c	d	174.	a	b	c	d
103.	a	b	c	d	139.	a	b	c	d	175.	a	b	c	d
104.	a	b	c	d	140.	a	b	c	d	176.	a	b	c	d
105.	a	b	c	d	141.	a	b	c	d	177.	a	b	c	d
106.	a	b	c	d	142.	a	b	c	d	178.	a	b	c	d
107.	a	b	c	d	143.	a	b	c	d	179.	a	b	c	d
108.	a	b	c	d	144.	a	b	c	d	180.	a	b	c	d
109.	a	b	c	d	145.	a	b	c	d	181.	a	b	c	d
110.	a	b	c	d	146.	a	b	c	d	182.	a	b	c	d
111.	a	b	c	d	147.	a	b	c	d	183.	a	b	c	d
112.	a	b	c	d	148.	a	b	c	d	184.	a	b	c	d
113.	a	b	c	d	149.	a	b	c	d	185.	a	b	c	d
114.	a	b	c	d	150.	a	b	c	d	186.	a	b	c	d
115.	a	b	c	d	151.	a	b	c	d	187.	a	b	c	d
116.	a	b	c	d	152.	a	b	c	d	188.	a	b	c	d
117.	a	b	c	d	153.	a	b	c	d	189.	a	b	c	d
118.	a	b	c	d	154.	a	b	c	d	190.	a	b	c	d
119.	a	b	c	d	155.	a	b	c	d	191.	a	b	c	d
120.	a	b	c	d	156.	a	b	c	d	192.	a	b	c	d
121.	a	b	c	d	157.	a	b	c	d	193.	a	b	c	d
122.	a	b	c	d	158.	a	b	c	d	194.	a	b	c	d
123.	a	b	c	d	159.	a	b	c	d	195.	a	b	c	d
124.	a	b	c	d	160.	a	b	c	d	196.	a	b	c	d
125.	a	b	c	d	161.	a	b	c	d	197.	a	b	c	d
126.	a	b	c	d	162.	a	b	c	d	198.	a	b	c	d
127.	a	b	c	d	163.	a	b	c	d	199.	a	b	c	d
128.	a	b	c	d	164.	a	b	c	d	200.	a	b	c	d
129.	a	b	c	d	165.	a	b	c	d					
130.	a	b	c	d	166.	a	b	c	d					
131.	a	b	c	d	167.	a	b	c	d					
132.	a	b	c	d	168.	a	b	c	d					
133.	a	b	c	d	169.	a	b	c	d					
134.	a	b	c	d	170.	a	b	c	d					
135.	a	b	c	d	171.	a	b	c	d					
136.	a	b	c	d	172.	a	b	c	d					

Patient Care and Education

Choose the best answer from the choices given.

1. Who is legally allowed to order an examination on a patient?
   a. nurse practitioner
   b. medical doctor
   c. technologist
   d. medical assistant

2. The unlawful touching of an individual without consent is considered which of the following?
   a. battery
   b. assault
   c. invasion of privacy
   d. tort

3. Parental informed consent is required on all patients under the age of
   a. 16.
   b. 17.
   c. 18.
   d. 21.

4. When correctly identifying a patient, which methods of verification should be used?
   1. Ask the patient his/her name.
   2. Check patient's wristband for accuracy.
   3. Check chart only.
   4. Find out date of birth.
   a. 1 only
   b. 1 and 2 only
   c. 3 only
   d. 1, 2, and 4 only

5. If a patient is unable to chew or cannot tolerate solid foods, which of the following is used to feed the patient?
   a. nasoenteric tube
   b. Swan Ganze catheter
   c. nasogastric tube
   d. central line

6. When performing a procedure in a sterile environment, the operator and surgeon should only pass each other in which direction?
   a. back to back
   b. front to back
   c. side to side
   d. back to front

7. The term *dyspnea* means
   a. difficulty sleeping.
   b. difficulty eating.
   c. difficulty breathing.
   d. difficulty seeing.

8. A sphygmomanometer is used to measure
   a. temperature.
   b. infection.
   c. pulse.
   d. blood pressure.

9. Nasocomial infections are acquired in which type of environment?
   a. school
   b. hospital
   c. doctor's office
   d. park

10. Which factors must be present to contribute to the spread of a disease?

    1. reservoir of infection
    2. pathogenic organism
    3. means of transmission
    4. susceptible host

  a. 1 and 2 only
  b. 1 and 4 only
  c. 3 and 4 only
  d. all of the above

11. Prior to the beginning and upon the end of a procedure, what is the minimal length of time you should wash your hands to be free of germs?
  a. 15 seconds
  b. 25 seconds
  c. 45 seconds
  d. 60 seconds

12. If a patient presents to the hospital with a life-threatening temperature above 105.8°F, what is this condition called?
  a. hypersthenic
  b. hyperglycemic
  c. hyperpyrexia
  d. hyperthemia

13. If your only method of sterilization is boiling, how long should you boil an item to make it usable in a sterile environment?
  a. 12 minutes
  b. 15 minutes
  c. 20 minutes
  d. 22 minutes

14. If a patient is placed on an airborne isolation, which of the following protective garment(s) must be worn prior to entering the room?

    1. mask
    2. gloves
    3. booties
    4. gown

  a. 1 and 2 only
  b. 1 and 4 only
  c. 1, 2, and 4 only
  d. all of the above

15. If a patient presents in the emergency room with multiple bee stings and seems to be going into shock, what type of shock is most associated with this situation?
  a. cardiogenic
  b. neurogenic
  c. septic
  d. anaphylactic

16. The term *hyperglycemia* refers to
  a. high blood sugar.
  b. high urine output.
  c. high creatinine levels.
  d. high blood pressure.

17. When scheduling patients for their procedures, when in the day would you schedule a barium enema?
  a. before noncontrast studies are performed
  b. mixed in with the noncontrast studies
  c. after noncontrast studies are performed
  d. the next day

18. When placing a patient on or off oxygen, what would be the normal flow rate to use?
   a. 2–4 L per minute
   b. 3–5 L per minute
   c. 1–3 L per hour
   d. 4–6 L per second

19. What is the main purpose of a chest tube?
   a. to remove fluid or air from the pleural space
   b. to supply a central line for IV injections
   c. to remove air or fluid from pleural space
   d. to feed patient

20. Air is considered which type of contrast agent?
   a. positive
   b. negative
   c. iodinated ionic
   d. iodinated nonionic

21. The acronym HIPAA stands for which of the following?
   a. Health Insurance Plan and Accountability Act
   b. Health Information Privacy and Accountability Act
   c. Health Information Portability and Accountability Act
   d. Health Insurance Portability and Accountability Act

22. Which of the following procedures require(s) an informed consent to be obtained prior to the start?
   1. upper GI series
   2. lumbar myelogram
   3. lower extremity angiogram

   a. 1 only
   b. 1 and 2 only
   c. 2 and 3 only
   d. 1, 2, and 3

23. Which of the following is (are) example(s) of symptoms involved with an anaphylactic reaction to contrast media?
   1. dyspnea
   2. hives
   3. vomiting

   a. 1 only
   b. 1 and 2 only
   c. 1 and 3 only
   d. 1, 2, and 3

24. Which is the most commonly used artery when obtaining a patient's blood pressure?
   a. carotid
   b. axillary
   c. coronary
   d. brachial

25. What is the condition termed when a resting heart rate is above 100 beats per minute?
   a. bradycardia
   b. tachycardia
   c. hypertension
   d. cardiomegaly

26. The most common means by which microorganisms are spread from person to person is
   a. dirty hands.
   b. contaminated clothing.
   c. soiled linen.
   d. radioactive materials.

Equipment Operation and Maintenance

Choose the best answer from the choices given.

27. Fluoroscopic examinations utilize a milliampere setting that is
    a. less than 5.
    b. stabilized between 10–15.
    c. stabilized between 20–30.
    d. variable.

28. The component of the x-ray tube which is a tungsten filament is called the
    a. anode.
    b. cathode.
    c. focusing cup.
    d. rotor.

29. A Compton's scattering reaction generally takes place between an incident photon and
    a. an inner shell target electron.
    b. an outer shell target electron.
    c. a target nucleus.
    d. a target proton.

30. A type of device which converts mechanical energy into electrical energy is called
    a. a cathode.
    b. an anode.
    c. a motor.
    d. a generator.

31. During bremsstrahlung radiation production, an incident electron
    a. strikes an outer shell electron of a target anode atom.
    b. strikes an inner shell electron of a target anode atom.
    c. breaks around the nucleus of a target atom and releases energy.
    d. loses all energy and ceases to exist.

32. The device during fluoroscopy that converts the attenuated x-ray photons into thousands of photons of light is the
    a. image intensifier.
    b. automatic brightness control.
    c. beam splitter.
    d. video monitor.

33. In an x-ray tube, if there are two filaments housed within the focusing cup, the tube is termed to be
    a. double filament.
    b. dual focus.
    c. bi-filament.
    d. dual wired.

34. Which of the following is (are) considered types of AEC?
    1. phototimer
    2. ionization chambers
    3. solid-state diodes

    a. 1 only
    b. 2 only
    c. 1 and 2 only
    d. 1, 2, and 3

35. As the target angle of the anode becomes steeper, what effect would this have on the actual focal spot?

a. becomes greater
b. becomes less
c. has no effect
d. actual focal spot will have heat load problems

36. Of which type of material(s) is the target track of the anode made from?
a. molybdenum
b. copper
c. aluminum
d. tungsten-rhenium

37. What is the main advantage of a rotating anode over a stationary anode?
a. a larger effective focal spot
b. higher quality electrons
c. better heat dissipation
d. better absorption of x-ray photons

38. The PBL of a collimator must be tested to be accurate to which percentage?
a. 2%
b. 5%
c. 10%
d. 15%

39. The mirror located within a collimator apparatus serves to do which of the following?

　　1. reflect light in order to reflect the path of the beam
　　2. work as inherent filtration
　　3. work as added filtration

a. 1 only
b. 1 and 2 only
c. 1 and 3 only
d. 1, 2, and 3

40. What is (are) the primary function(s) of the automatic brightness control during fluoroscopy?

　　1. to vary the mAs and kVp for part thickness
　　2. to adjust the amplification of the flux gain
　　3. to convert a greater degree of electrons into light photons

a. 1 only
b. 2 only
c. 3 only
d. 1, 2, and 3

41. The product of minification gain and flux gain describes
a. total brightness gain.
b. grid ratio.
c. magnification factor.
d. total filtration.

42. The area of the image intensifier in which light strikes and gets converted into an electron "image" is called the
a. input phosphor.
b. output phosphor.
c. focusing lens.
d. photocathode.

43. The percentage of characteristic production in an x-ray tube is generally around
a. 0–10%.
b. 15–20%.
c. 50%.
d. 80–85%.

44. The potential difference between two points within a circuit describes
a. amperage.
b. current.
c. voltage.
d. resistance.

45. The degree of resolution in computed radiography will increase as
  a. photostimulable phosphor size increases.
  b. helium neon laser beam size increases.
  c. monitor matrix size increases.
  d. mAs increases.

46. The greatest advantage of computed radiography over analog radiography is
  a. cost.
  b. exposure latitude.
  c. more repeated exposures.
  d. less cassette storage.

47. Which of the following is required for the creation of x-ray photons?

  1. high speed electrons striking a target disc
  2. electrons causing a breaking reaction around a target nucleus
  3. electrons reacting with an outer shell target electron

  a. 1 only
  b. 1 and 2 only
  c. 2 and 3 only
  d. 1, 2, and 3

48. The dielectric oil which surrounds and helps to cool the x-ray tube also serves as a type of x-ray filtration. This type of filtration is commonly referred to as
  a. inherent filtration.
  b. added filtration.
  c. bremsstrahlung filtration.
  d. compound filtration.

49. Which of the following is found in the filament circuit and functions to vary resistance?
  a. step-up transformer
  b. autotransformer
  c. rheostat
  d. capacitor

50. Which of the following is NOT true about collimators?
  a. Infinite field sizes are available.
  b. They allow better filtration of the x-ray beam.
  c. They allow better cleanup of scattered radiation.
  d. They provide a light field for the path of the x-ray beam.

Image Production and Evaluation

Choose the best answer from the choices given.

51. If a film is mishandled prior to processing, it is likely that the film will exhibit areas of
  a. tree static artifact.
  b. starry night artifact.
  c. minus density artifact.
  d. plus density artifact.

52. Given the following groups of exposure factors, which will exhibit the greatest density on a film?
  a. 300 mA, 0.08 sec, 8:1 grid ratio, 72 inches
  b. 200 mA, 0.25 sec, 8:1 grid ratio, 72 inches
  c. 300 mA, 0.08 sec, 8:1 grid ratio, 40 inches
  d. 300 mA, 0.08 sec, 8:1 grid ratio, 36 inches

53. In an automatic processor, the function of the developer is to
  a. remove all unexposed silver halide crystals.
  b. reduce the manifest image into a latent image.
  c. wash the film.
  d. serve as the sensitivity specks.

54. Which of the following does NOT contribute to radiographic contrast on an image?
a. atomic number of target
b. kilovoltage used
c. mAs used
d. use of a radiographic grid

55. Which of the following would NOT be a reason to utilize a high ratio grid?
a. absorption of scattered radiation
b. when a low-dose technique is required
c. to present a short scale of contrast
d. for an extremely thick body part

56. In the characteristic curve of a specific type of film, what does the toe portion represent?
a. overexposure
b. underexposure
c. emission spectrum
d. no exposure

57. Which of the following factor(s) is (are) necessary in order for quantum mottle to be present on an image?

1. fast speed screen
2. high kVp technique
3. low mAs technique

a. 1 only
b. 1 and 2 only
c. 1 and 3 only
d. 1, 2, and 3

58. Which of the following exposure techniques will produce the shortest scale of contrast?
a. 200 mA, 0.75 sec, 52 kVp
b. 400 mA, 0.60 sec, 58 kVp
c. 300 mA, 0.10 sec, 65 kVp
d. 800 mA, 0.05 sec, 90 kVp

59. Magnification and foreshortening are considered
a. size distortion.
b. shape distortion.
c. film distortion.
d. average gradient.

60. The source-to-image distance and the size of the focal spot have the greatest effect on which radiographic property?
a. contrast
b. density
c. recorded detail
d. shape distortion

61. The overall blackening present on a radiograph is the definition of
a. density.
b. contrast.
c. magnification.
d. emission spectrum.

62. The difference between adjacent densities present on a radiograph is the definition of radiographic
a. density.
b. contrast.
c. recorded detail.
d. filtration.

63. Which of the following groups of exposure factors would produce the greatest radiographic density?
a. 10 mAs, 60 kVp, 8:1 grid ratio
b. 20 mAs, 60 kVp, 16:1 grid ratio
c. 10 mAs, 74 kVp, no grid
d. 10 mAs, 74 kVp, 6:1 grid

64. If a 50-inch SID is utilized on an object located 5 inches from the image receptor, what will the magnification factor be?
a. 1
b. 1.1
c. 1.5
d. 2

65. What apparatus could be utilized when imaging a body part with varying thicknesses?
a. grid
b. compensating filter
c. inherent filter
d. slow speed screen/film

66. What is the primary advantage of a moving grid over a stationary grid?
a. blurring of the grid strips
b. better scatter cleanup
c. ease of portable use
d. better cleanup of primary beam

67. Which of the following factors would produce the film with the longest scale of contrast?
a. 200 mA, 0.1 sec, 70 kVp, 6:1 grid ratio
b. 100 mA, 0.2 sec, 70 kVp, 8:1 grid ratio
c. 200 mA, .01 sec, 70 kVp, 12:1 grid ratio
d. 100 mA, 0.2 sec, 70 kVp, 16:1 grid ratio

68. When a singular mA is used, regardless of the combination of mA and time that is used in order to produce it, the resulting images will produce the same density. This describes the principal of
a. the inverse square law.
b. the density maintenance law.
c. the 15% rule.
d. reciprocity.

69. Films that are pulled from the film bin quickly in humid conditions run the risk of producing
a. fog.
b. minus density.
c. static artifact.
d. pick-off.

70. Any material that emits light upon stimulation from a type of radiating photon is called
a. an electron.
b. a phosphor.
c. an image receptor.
d. a filter.

71. Which of the following factors can be used in order to regulate radiographic density?
  1. exposure time
  2. milliamperage
  3. kilovoltage

a. 2 only
b. 1 and 2 only
c. 2 and 3 only
d. 1, 2, and 3

72. How much exposure time would be necessary if 400 mA was utilized in order to produce 20 mAs?
a. 0.025 sec
b. 0.05 sec
c. 0.10 sec
d. 0.50 sec

73. An original exposure of 200 mA, 0.25 sec, and 90 kVp was utilized; in order to minimize motion artifact on the image, a technique utilizing 400 mA, 90 kVP, and _____ can be used.
a. 0.5 sec
b. 0.25 sec
c. 0.125 sec
d. 0.0625 sec

74. What would the mA be if 0.25 seconds was used in order to produce 100 mAs?
   a. 200 mA
   b. 400 mA
   c. 600 mA
   d. 1,000 mA

75. Which of the following factors may contribute to an increase in recorded detail on the finished image?

       1. large focal spot
       2. smaller focal spot
       3. slow screen film

   a. 1 only
   b. 2 only
   c. 2 and 3 only
   d. 1 and 3 only

76. If 20 mAs were utilized for a particular exposure utilizing a par (200) speed screen, what would the new mAs be if a rare earth (800) speed screen was utilized?
   a. 5
   b. 10
   c. 15
   d. 20

77. The terminology short-scale contrast generally refers to a film which contains
   a. few densities with great differences between them.
   b. a large amount of densities with fewer differences between them.
   c. the same density present throughout the entire image.
   d. no density present throughout the entire image.

78. In order to properly test screen/film contact, which type of exam is performed?
   a. spin top
   b. dosimetry
   c. wire mesh
   d. kV calibration

79. In an automatic processor, what is the purpose of the crossover rollers?
   a. to direct a film from the bottom of the solution tank back toward the top
   b. to direct a film into the processor
   c. to direct a film though the dryer
   d. to direct a film from one solution into the next

80. If the kilovoltage had to be increased in order to assist in the reduction of patient dosage, the most noticeable effect on the finished image would be
   a. decreased contrast.
   b. increased contrast.
   c. increased light fog.
   d. decreased density.

81. The height of a grid's lead strips to the interspace distance describes
   a. grid frequency.
   b. grid latitude.
   c. grid ratio.
   d. grid cutoff.

82. When scattered radiation strikes a film, the end result on the image is which of the following?

    1. increased fog
    2. decreased density
    3. decreased contrast

  a. 1 only
  b. 1 and 2 only
  c. 1 and 3 only
  d. 1, 2, and 3

83. An unexposed film exhibits some density due to
  a. blue dye.
  b. bin fog.
  c. light fog.
  d. radiation fog.

84. Short-scale contrast is a product of
  a. high mAs.
  b. low mAs.
  c. high kVp.
  d. low kVp.

85. As the speed of film increases, if utilizing the same technique as a slow speed film, what would happen to the density on the film?
  a. It would increase.
  b. It would decrease.
  c. It would remain the same.
  d. It would be unnoticeable.

86. In an exposure, filtration serves what main purpose?
  a. It reduces technologist dosage.
  b. It increases patient dosage.
  c. It decreases patient dosage.
  d. It reduces scattered radiation production.

87. If 32 mAs and 70 kVp are utilized for a given exposure using an 8:1 grid, if the exposure was to be repeated using a 16:1 grid, what would the new mAs be if all other factors remain the same?
  a. 16 mAs
  b. 32 mAs
  c. 50 mAs
  d. 64 mAs

88. A perceptible increase in radiographic density would be noticed only if mAs were increased
  a. 10%.
  b. 15%.
  c. 20%.
  d. 30%.

89. A film was produced utilizing 60 kVp and 10 mAs. In order to decrease dosage to the patient, the film could be repeated utilizing which technique in order to maintain density?
  a. 60 kVp, 5 mAs
  b. 70 kVp, 5 mAs
  c. 50 kVp, 20 mAs
  d. 70 kVp, 10 mAs

90. Scattered radiation is produced
  a. at the x-ray tube.
  b. at the cassette.
  c. at the tabletop.
  d. in the patient.

91. Which of the following will not affect patient dose?
  a. milliamperage per second
  b. filtration
  c. focal spot size
  d. kilovoltage peak

92. During automatic processing, the replenishment rate of the chemicals associated is determined by which factor?
a. size of film
b. temperature of processor
c. speed of the roller system
d. type of chemicals used

93. The term *quantity*, when pertaining to the x-ray beam, commonly refers to which technical factor?
a. mAs
b. kVp
c. grid ratio
d. object-to-image distance

94. In relation to recorded detail, what effect would an increase in the object-to-image distance cause to the image?
a. Detail would not be affected.
b. Detail would decrease inversely.
c. Detail would increase proportionally.
d. Detail would increase inversely.

95. In automatic processors, what is the primary purpose of the rollers?
a. transport film throughout processor
b. recirculate chemistry throughout
c. heat up the film sufficiently to assist in drying
d. collect excess, unexposed black metallic silver

96. Which would the effect be in an image if there was an increase in the photon energy in an x-ray beam?
a. increased contrast
b. decreased contrast
c. decreased density
d. no effect

97. An image of the knee was obtained utilizing 10 mAs, 60 kVP, and an 8:1 grid with 40 inches of source-to-image distance. If the image was now to be obtained utilizing 20 mAs, with all other factors remaining the same, the end result would show
a. increased density.
b. decreased density.
c. decreased contrast.
d. increased magnification.

98. Which of the following could lead to a film emerging from a processor to appear underexposed?
a. processor not sufficiently warmed up
b. developer temperature too hot
c. excessive kVp
d. excessive dryer temperature

99. In order to produce a change equaling double the original density on an image, the kVp must be manipulated at least
a. 5%.
b. 15%.
c. 20%.
d. 30%.

100. Which of the following will allow for elongation of a part on an image?
  1. angling the x-ray tube to the part
  2. angling of the tube to the IR
  3. angling of the grid to the IR

a. 1 only
b. 2 only
c. 3 only
d. 1, 2, and 3

Radiographic Procedures

Choose the best answer from the choices given.

101. A type of radiological exam in which motion generated by both the x-ray tube and the film used to image a specific plane in the body is commonly referred to as
   a. magnetic resonance imaging.
   b. Hida scan.
   c. computed tomography.
   d. general conventional tomography.

102. When imaging the skull in the lateral position, which of the following baselines must be utilized?
   a. IOML
   b. MSP
   c. IPL
   d. all of the above

103. For the AP projection of the lower axillary ribs, the IR is aligned so that the lower border is at the level of which of the following?
   a. symphysis pubis
   b. iliac crest
   c. ASIS
   d. one inch above the shoulder

104. Which anatomical structure is best demonstrated in the AP open mouth projection of the cervical spine?
   a. C-3
   b. atlas
   c. dens
   d. vertebra prominens

105. When performing a left lateral decubitis of the abdomen during a double contrast barium enema, which structure is best shown?
   a. lateral wall of the ascending colon
   b. rectum
   c. ileocecal valve
   d. lateral wall of descending colon

106. Which of the following anatomical structures is best visualized during an AP projection of the hip with the lower leg internally rotated?
   a. lesser tuberosity
   b. femoral neck
   c. ischial tuberosity
   d. symphysis pubis

107. The carpal scaphoid is best visualized utilizing which projection?
   a. Gaynor-Hart Projection
   b. AP projection of the carpal bones
   c. PA projection of the carpal bones with ulnar flexion
   d. PA projection of the carpal bones with radial flexion

108. Which landmark is located at the level of T-2/T-3?
   a. scapular notch
   b. sternal notch
   c. vertebra prominens
   d. cauda equina

109. For accurate imaging of the paranasal sinuses, the patient must be in which position?
   a. recumbent
   b. lateral decubitis
   c. prone
   d. erect

110. When performing a lateral projection of the cranium, the central ray is directed perpendicular to the plane of the IR entering at which point?
    a. two inches inferior to EAM
    b. two inches superior to EAM
    c. one and a half inches posterior to the upper ear attachment
    d. outer canthus

111. Which of the following structures is located in the right upper quadrant of the abdomen?
    a. liver
    b. appendix
    c. splenic flexure
    d. ileocecal valve

112. Which of the following describes a patient lying down with his or her feet elevated and the upper body and head lower?
    a. Fowler
    b. supine
    c. Sims
    d. Trendelenburg

113. When performing an oblique projection of the lumbar spine, the eye of the "Scottie dog" represents which of the following structures?
    a. pedicle
    b. lamina
    c. superior articulating process
    d. inferior articulating process

114. The proper degree of obliquity for an AP oblique projection of the right SI joint is
    a. 10 degrees.
    b. 25 degrees.
    c. 35 degrees.
    d. 45 degrees.

115. A hangman's fracture refers to a fracture of which bony structure?
    a. symphysis pubis
    b. C-7
    c. dens
    d. gonion

116. Which view best demonstrates the ankle mortise?
    a. AP projection of ankle
    b. AP 45 degree oblique projection of ankle
    c. AP 15 degree oblique projection of ankle
    d. lateral projection of ankle

117. Which of the following is considered involuntary motion?
    a. breathing
    b. flexing the elbow
    c. hyperextension of the neck
    d. heartbeat

118. Which position best demonstrates the carpal pisiform in profile?
    a. ulnar flexion
    a. AP projection of wrist
    c. PA oblique of wrist
    d. AP oblique of wrist

119. Which organ is responsible for the production of bile?
    a. spleen
    b. liver
    c. pancreas
    d. gallbladder

120. Which organ is responsible for the regulation of insulin within the body?
    a. liver
    b. gallbladder
    c. pancreas
    d. stomach

121. At which of the following organs is peristalsis considered the slowest?
    a. esophagus
    b. colon
    c. stomach
    d. small intestine

122. The junction between the stomach and small intestine is called the
    a. cardiac orifice.
    b. pyloric valve.
    c. gastroesophageal junction.
    d. mitral valve.

123. In order to best visualize a left-sided pneumo-thorax in a patient who cannot assume the erect position, what would be the best alternative?
    a. right lateral decubitis
    b. left lateral decubitis
    c. supine
    d. prone

124. For a properly visualized KUB projection, the lower margin of the cassette should be at the level of
    a. iliac crest.
    b. symphysis pubis.
    c. xiphoid tip.
    d. ASIS.

125. When a patient is supine with IOML perpen-dicular to the plane of the table, and the central ray is directed 37 degrees caudally entering two inches above the glabella, the structure best visualized is the
    a. frontal sinus.
    b. petrous pyramids.
    c. sphenoid bone.
    d. occipital bone.

126. The small bowel series is generally completed when the radio-opaque contrast reaches the
    a. rectum.
    b. ileocecal valve.
    c. GE junction.
    d. sigmoid.

127. When a body part moves toward the midline of the body, the term that is   generally used is
    a. abduction.
    b. adduction.
    c. proximal.
    d. Trendelenberg.

128. The sternum is generally positioned in a shal-low RAO position in order to
    a. bring it more in line with the ribs.
    b. clear it from any obstruction.
    c. use the heart shadow as contrast.
    d. make it more comfortable for the patient.

129. The raised surface on the proximal anterior portion of the tibia is called the
    a. tibial plateau.
    b. tibial tuberosity.
    c. tibial ridge.
    d. tibial condyle.

130. In order to best visualize the glenoid cavity in profile, the patient should be positioned in
    a. an AP position.
    b. a PA position.
    c. an AP 45 degree oblique position.
    d. a PA 60 degree oblique position.

131. In the case where there are suspected bilateral fractures of the femoral necks, which of the following positions should be imaged?

    1. AP

    2. Danelius-Miller

    3. Clements-Nakayama

  a. 1 only

  b. 1 and 2 only

  c. 1 and 3 only

  d. 1, 2, and 3

132. In order to best visualize the AC joints bilaterally, the exam must best be performed in which of the following positions?

    1. supine

    2. with weights

    3. without weights

  a. 1 only

  b. 1 and 2 only

  c. 2 and 3 only

  d. 1, 2, and 3

133. There are ___ carpal bones in both wrists.

  a. 4

  b. 8

  c. 14

  d. 16

134. Facial bone studies should always be performed in which position?

  a. supine

  b. prone

  c. erect

  d. lateral decubitis

135. The thoracic spine presents which type of curve?

  a. lateral

  b. lordotic

  c. oblique

  d. kyphotic

136. When performing projections of the right sacroiliac joint, the patient can be positioned in which of the follwing positions?

    1. LPO

    2. RPO

    3. RAO

  a. 1 only

  b. 1 and 3

  c. 2 and 3

  d. 1, 2, and 3

137. Which projection of the cervical vertebra best demonstrates the zygapophyseal joints?

  a. RAO

  b. LAO

  c. AP

  d. lateral

138. Which of the following projections is best utilized in order to rule out a blowout fracture?

  a. Leonard George

  b. Haas

  c. Rheese

  d. Stenvers

139. The carpal interspaces are best visualized utilizing the

  a. AP position.

  b. PA position.

  c. lateral position.

  d. AP oblique position.

140. For the proper visualization of the Danelius-Miller projection, the cassette must be placed in an upright device
   a. perpendicular to the femoral neck.
   b. parallel with the femoral neck.
   c. parallel with the iliac crest.
   d. perpendicular to the iliac crest.

141. Which vertebra contains a foramen throughout its transverse processes?
   a. cervical
   b. thoracic
   c. lumbar
   d. sacrum

142. The line drawn between the outer canthus of the eye and the external auditory meatus, which assists in the positioning of the skull, is commonly referred to as
   a. AML.
   b. GAL.
   c. OML.
   d. IOML.

143. Which projection of the shoulder presents the greater tubercle of the humerus in profile on the finished image?
   a. axillary projection
   b. anteroposterior projection internal rotation
   c. anteroposterior projection external rotation
   d. scapula "Y" projection

144. The tarsal talus articulates with which bone inferiorly?
   a. navicular
   b. calcaneous
   c. cuboid
   d. metatarsal

145. When performing a projection of the ribs, in order to best demonstrate the lower ribs below the diaphragm, the technologist should instruct the patient to
   a. suspend breathing.
   b. suspend breathing at the end of full inspiration.
   c. suspend breathing at the end of full exhalation.
   d. utilize a breathing technique.

146. Which is the most frequently fractured carpal bone?
   a. capitate
   b. navicular
   c. pisiform
   d. hamate

147. During a lateral projection of the knee, in order to elevate the medial condyle of the femur out of the joint space, the technologist should
   a. utilize a perpendicular central ray.
   b. angle the central ray 5 degrees cephalic.
   c. angle the central ray 5 degrees caudad.
   d. flex the knee 60–90 degrees.

148. In the modified Waters Projection for sinuses, which set of sinuses are projected through the open mouth?
   a. frontal
   b. ethmoid
   c. sphenoid
   d. maxillary

149. When performing an AP projection of the sacrum, to compensate for the normal degree of kyphosis, the central ray should be directed
   a. 35 degrees caudad.
   b. 15 degrees cephalic.
   c. 10 degrees caudad.
   d. 25 degrees cephalic.

150. How many posterior ribs should be visualized above the diaphragm on a properly exposed PA chest radiograph?
    a. 6
    b. 8
    c. 10
    d. 12

151. The cranium is composed of how many bones?
    a. 4
    b. 6
    c. 8
    d. 10

152. The spinal cord terminates at which vertebral level?
    a. T-11/T-12
    b. S-1/S-2
    c. C-7/T-1
    d. L-1/L-2

153. The right kidney generally lies on a lower plane than the left kidney. This is due to the presence of which organ?
    a. liver
    b. colon
    c. heart
    d. pancreas

154. The largest bone in the upper extremity is the
    a. femur.
    b. capitate.
    c. humerus.
    d. ulna.

155. The pituitary gland lies in which structure in the cranium?
    a. sella turcica
    b. mastoid process
    c. pterygoid process
    d. bregma

156. An asthenic body habitus refers to a person who
    a. is extremely thin with a narrow and shallow thorax.
    b. is slender and light in weight.
    c. has an athletic build.
    d. has a massive build with a deep and broad thorax.

157. The scapular "Y" projection describes a PA oblique position with the body rotated approximately how many degrees from the plane of the IR?
    a. 30
    b. 45
    c. 60
    d. 90

158. For a PA projection of the hand, the central ray should be directed
    a. to the third metatarsophalangeal joint.
    b. to the second metacarpophalangeal joint.
    c. to the third metacarpophalangeal joint.
    d. to the midline of the proximal row of carpals.

159. When performing a PA projection of the wrist, in order to reduce the natural OID between the carpals and the plane of the IR, the technologist could
    a. angle the CR 5 degrees caudally.
    b. perform radial flexion.
    c. place the shoulder on the same plane.
    d. flex the digits of the hand.

160. In the case where a patient cannot extend the elbow joint, in order to perform a thorough study, the technologist should
    a. force flexion.
    b. perform two AP projections.
    c. send the patient away.
    d. perform only a lateral projection.

Radiation Protection

Choose the best answer from the choices given.

161. The effects of radiation are known to be influenced by all of the following EXCEPT
   a. dose rate.
   b. size of the cell(s).
   c. type of cell.
   d. type of radiation.

162. The effects of radiation are usually more profound if
   a. the exposure is delivered in one dose.
   b. the exposure is delivered in small multiple doses.
   c. the exposure is delivered to a small area.
   d. the exposure is delivered with a filter.

163. Shorter exposure times, decreased patient dose, and increased latitude in technique are advantages of
   a. high kV techniques.
   b. high contrast techniques.
   c. high grid ratio techniques.
   d. high mAs techniques.

164. Radiation amounts received during diagnostic studies are chiefly affected by
   a. focal spot size.
   b. patient positioning.
   c. size of field exposed.
   d. type of equipment.

165. Which of the interactions are most common in tissue at 100 kVp?
   a. coherent
   b. pair production
   c. Compton scatter
   d. photoelectric effect

166. General purpose radiography units' minimal filtration thickness is
   a. 0.5 mm Al eq.
   b. 1.75 mm Al eq.
   c. 2.5 mm Al eq.
   d. 4.0 mm Al eq.

167. What is the shortest distance from the tube to the patient when employing a C-arm unit?
   a. 6 inches
   b. 9 inches
   c. 12 inches
   d. 18 inches

168. What unit is ionization in air measured in?
   a. gray
   b. C/kg
   c. curie
   d. ievert

169. Which of the following is the safest time to radiograph a fetus?
   a. third trimester
   b. second trimester
   c. first trimester
   d. none of the above

170. Any added filtration in a radiography unit should eliminate most
   a. primary radiation.
   b. shorter wavelengths.
   c. remnant radiation.
   d. longer wavelengths.

171. Which of the following is a property of radiation that accounts for its effect on the biological systems?
   a. Radiation has different energies.
   b. Radiation has no electric charge.
   c. Radiation causes ionization in matter.
   d. Radiation travels at the speed of light.

172. The degree to which recovery is possible after exposure to radiation is related to
   a. total dose received.
   b. degree of protraction.
   c. LET of the radiation.
   d. all of the above.

173. A piece of paper will provide adequate protection from which of the following types of radiation?
   a. alpha
   b. beta
   c. gamma
   d. x-rays

174. LET stands for
   a. lethal effective tolerance.
   b. lethal exchange table.
   c. linear energy transmutation.
   d. linear energy transfer.

175. Human cells are divided into two classifications. There are germ cells and
   a. organic cells.
   b. benign cells.
   c. somatic cells.
   d. reproductive cells.

176. What is the process of cell division of somatic cells called?
   a. meiosis
   b. mitosis
   c. synthesis
   d. multiplication

177. The direct hit theory of cell irradiation can be described by which of the following statements?
   a. The DNA molecule is struck.
   b. The cell nucleus is struck.
   c. The cell cytoplasm is struck.
   d. The cell is ionized.

178. Which of the following would describe the shape of a DNA molecule?
   a. elliptical
   b. hypocycloidal
   c. double helix
   d. circular

179. Which of the following is a type of damage that can occur to a DNA molecule when exposed to radiation?
   a. change in the genetic code
   b. breakage of the chromosomes
   c. breakage of the DNA molecule
   d. all of the above

180. What do the letters DNA stand for?
   a. direct nuclear action
   b. deuteron nucleus activity
   c. deoxyribonucleic acid
   d. deoxyribose neutral acid

181. How will oxygen retention affect the radiosensitivity of a cell?
   a. Radiosensitivity will increase.
   b. Radiosensitivity will decrease.
   c. Radiosensitivity will be eliminated.
   d. Radiosensitivity will not be affected.

182. Irradiation of which of the following anatomical areas will affect the production of white blood cells?
   a. lungs
   b. liver
   c. spleen
   d. bone marrow

183. What do the letters RBE stand for?
a. relative biologic effectiveness
b. radiosensitive biologic effect
c. radioactive biological energy
d. radiation bypass effect

184. Which of the following demonstrates that distance is one of the three cardinal principles of radiation protection?
a. Law of Bergonie and Tribondeau
b. inverse square law
c. radiation power formula
d. grid conversion formula

185. Which term describes the separation of water into hydrogen and oxygen due to irradiation?
a. radioactivation
b. radiolysis
c. interaction
d. toxicity

186. What is meant by the term *interphase death* in radiobiology?
a. Cells die before going into interphase.
b. Cells die before leaving interphase.
c. Cells die in between mitotic phases.
d. The organism dies before the average cell leaves interphase.

187. What term describes cell damage from radiation that does not kill the cell?
a. sublethal death
b. cell repair
c. cell healing
d. cognition

188. Adding or increasing filtration will primarily protect which of the following?
a. radiographer
b. reproductive organs
c. patient's skin
d. patient's bone marrow

189. What should you do if preparing to shield a patient for fluoro?
a. use a shadow shield
b. place the shield below the patient
c. place the shield on top of the patient
d. none of the above

190. All are major steps in acute radiation syndrome EXCEPT
a. latent period.
b. manifest illness.
c. prodromal stage.
d. dormant stage.

191. All of the following are effects from early (acute) somatic syndrome EXCEPT
a. erythemia.
b. cancer.
c. epilation.
d. desquamation.

192. Which of the following adhere to the main principals of radiation protection?

1. increase in mA
2. decrease in source-to-image distance
3. increase in source-to-image distance
4. increase in shielding

a. 1 and 2 only
b. 3 and 4 only
c. 2 and 4 only
d. all of the above

193. A wall which serves as a primary radiation barrier must be minimally how high?
a. 6 feet
b. 7 feet
c. 8 feet
d. 9 feet

194. During a fluoroscopic procedure, a radiation monitor should be worn at the
   a. collar, outside the lead apron.
   b. waist, inside the lead apron.
   c. collar, inside the lead apron.
   d. waist, outside the lead apron.

195. The total occupational dose limit allowable for a radiographer per year should NOT exceed
   a. 5 rads.
   b. 5 rems.
   c. 5 grays.
   d. 5 curies.

196. If an occupational worker receives 38 R/min working 10 feet from the source, how many R/min will the person receive if he or she moved 20 feet from the source?
   a. 2.1 R/min
   b. 9.5 R/min
   c. 21.1 R/min
   d. 152 R/min

197. Within an x-ray room, the amount of radiation present during fluoroscopic exposure is measured in
   a. rad.
   b. rem.
   c. roentgen.
   d. ohm.

198. The term ALARA refers to
   a. as long as reasonably achievable.
   b. as low as reasonably achievable.
   c. as light as reasonably allowable.
   d. as low as realistically available.

199. When utilizing gonadal shielding, the exposure to a patient's ovaries will be reduced up to what percentage?
   a. 5%
   b. 25%
   c. 75%
   d. 95%

200. In which of the following circumstances is it possible for a technologist to hold a patient during a radiographic procedure?
   a. during trauma radiography
   b. when radiographing a child under the age of three
   c. when a patient is combative
   d. none of the above

Radiography Practice Test II Answers

Patient Care and Education

1. **b.** Only an MD is allowed to request/order an examination.
2. **a.** Unlawfully touching an individual without his or her consent is battery and is punishable in a criminal court.
3. **c.** Any person(s) under the age of 18 years old is considered a minor and cannot sign a legal document.
4. **b.** When verifying a patient, it is always critical to have the patient tell you his or her name and date of birth as well as check his or her wristband.
5. **c.** A nasogastric tube is used to feed or suction the stomach.
6. **a.** The operator and surgeon should only pass back to back in a sterile environment to avoid contamination.
7. **c.** Dyspnea is when a patient is having difficulty breathing.
8. **d.** A patient's blood pressure is measured by using a sphygmomanometer.
9. **b.** Nasocomial infections are only hospital acquired.
10. **d.** All of the above must be present to form the cycle of disease.
11. **a.** You must wash your hands for a minimum of 15 seconds to be germ-free.
12. **c.** Hyperpyrexia is the life-threatening temperature above 105.8°F.
13. **a.** When boiling an item for sterilization, each item must be boiled for a minimum of 12 minutes.
14. **c.** Airborne isolation requires all technologists entering the patient's room to wear a mask and gown for protective purposes. Gloves are a universal precaution and are required for all examinations.
15. **d.** Anaphylactic shock is cause by the injection of a foreign protein such as a bee sting or contrast agent.
16. **a.** Hyperglycemia is a term commonly used for high blood sugar.
17. **c.** Due to the length of a barium enema study, it is always wise to schedule a case after all noncontrast studies have been performed to avoid delays.
18. **b.** 3–5 L per minute is the normal flow rate when placing a patient on oxygen, unless otherwise instructed by a physician.
19. **c.** The main function of a chest tube is to remove air and fluid from the pleural space.
20. **a.** Air is considered to be a positive contrast agent. A great example of positive contrast is a chest radiograph.
21. **d.** The Health Insurance Portability and Accountability Act established legal guidelines pertaining to patient information and medical records.
22. **c.** All procedures requiring intervention require informed patient consent prior to the start. Examples of such studies include angiography, myelography, and embolizations.
23. **d.** Dyspnea, hives, and vomiting are all reactions to anaphalaxis.
24. **d.** The brachial artery located at the elbow is the most frequent utilized artery when obtaining blood pressure.
25. **b.** The condition of tachycardia is when the heart beats more than 100 beats/minute at rest.
26. **a.** Soiled hands are the most common means of transmitting microorganisms.

Equipment Operation and Maintenance

27. **a.** Fluoroscopic equipment utilizes a mA setting that is less than 5 mA.
28. **b.** The tungsten filament which when heated emits boiled-off electrons for the use of x-ray production is called the cathode.

29. b. Compton's scattered photons exist after an incident electron strikes an outer shell target electron, ejecting the electron and changing the path of the incident photon.

30. d. A generator is defined as an object which converts mechanical energy into electrical energy.

31. c. A bremsstrahlung reaction takes place when an incident electron within the x-ray tube breaks around a target anode atom, losing energy as it slows down. The energy is released as an x-ray photon.

32. a. The image intensifier serves to convert attenuated photons emerging from the patient into photons of light in order to create the image.

33. b. When two filaments are housed in one focusing cup of an x-ray tube, the tube is called dual focus.

34. c. Both phototimers and ionization chambers are examples of AEC units.

35. a. As the target angle of the anode becomes steeper, the actual focal spot becomes greater. It still allows for small focal spots and also has a greater heat load capacity. However, it will have a more accentuated anode heel effect.

36. d. The anode target track is made up of a tungsten-rhenium compound. The disc is made up of molybdenum.

37. c. The major advantage of a rotating anode is better heat dissipation, which allows for greater techniques to be utilized.

38. a. The PBL must be no greater than 2% accurate between the light path and the path of the primary beam.

39. c. The mirror located in the collimator box serves as added filtration, and it helps reflect the light from a bulb to mirror the path of the primary beam.

40. a. The automatic brightness control serves to adjust the technique during fluoroscopy in order to maintain brightness when imaging varying part thicknesses.

41. a. Minification gain × flux gain = total brightness gain.

42. d. The photocathode is the part of the image intensifier connected to the input phosphor that converts the light generated by the input phosphor into electrons to be focused across to the output phosphor.

43. d. Characteristic interactions within the x-ray tube account for 80–85% of the interactions during x-ray production.

44. c. Voltage is defined as a potential difference between two points.

45. c. Increasing monitor matrix size, decreasing laser size, and decreasing photostimulable phosphor size all lead to increases in resolution in computed radiography.

46. b. The greatest advantage of CR over analog imaging is the latitude given per exposure, which in turn helps reduce patient dosage.

47. b. X-ray photons are produced in an x-ray tube from a combination of high-speed electrons causing either bremsstrahlung (or breaking) reactions or inner shell reactions (characteristic) at the anode target.

48. a. Inherent filtration is any filtration that is found locally at the x-ray tube itself. Examples of such are the glass envelope, the dielectric oil, etc.

49. c. A rheostat functions to vary resistance.

50. c. A collimator serves no purpose in the cleanup of produced scattered radiation.

Image Production and Evaluation

51. c. When a film is mishandled, areas of minus (or lack of) density are present at the areas of mishandling.

52. d. With the given techniques, the closest distance will exhibit the greatest density.

53. b. The developer solution serves to reduce the manifest image into a visible latent image by chemical interaction with the film.

54. **c.** The mAs effect on radiographic contrast is negligible.

55. **b.** The higher the ratio of grid utilized, the greater the exposure required in order to visualize the part, thus increasing patient dosage.

56. **b.** The toe on the characteristic curve represents underexposure.

57. **d.** Fast speed film, high kilovoltage, and low mAs techniques all lead to quantum mottle, which is visualized as a speckled appearance on a processed image.

58. **a.** The lower the kilovotage, the shorter scale of contrast that could be achieved.

59. **a.** Magnification and foreshortening are considered size distortion.

60. **c.** SID, OID, and focal spot size all have a pronounced effect on recorded detail.

61. **a.** Density is defined as the overall blackening on a radiograph.

62. **b.** Contrast is defined as the differences between adjacent densities on a radiograph.

63. **c.** 10 mAs, 74 kVp, and no grid would produce the greatest density on an image given these techniques.

64. **b.** SID/SOD = MF, 50/45 = 1.1.

65. **b.** The compensating filter is utilized in order to produce a more uniform contrast when varying part thicknesses are present on a radiograph.

66. **a.** The primary purpose of grid movement during exposure is to blur the lead strips present in the grid to lessen their visibility.

67. **a.** The smaller the grid ratio, the longer scale of contrast present on a radiograph.

68. **d.** The law of reciprocity is defined as any mAs value that can be attained utilizing any combination of mA and time which will generate the same density.

69. **c.** Static artifact is generally caused by low humid conditions combined with some mishandling.

70. **b.** A phosphor is any material which emits light after stimulation by radiation.

71. **d.** Milliamperage, kilovotage (15% rule), and exposure time can all be manipulated to affect density.

72. **b.** mAs = mA × time, 20 mAs = 400 mA × 0.05 sec.

73. **c.** mAs = mA × time, 50 mAs = 400 mA × 0.125 sec.

74. **b.** mAs = mA × time, 100 mAs = 400 mA × 0.25 sec.

75. **c.** Small focal spot sizes as well as slower film/screen speeds both contribute to increasing radiographic detail.

76. **a.** new mAs = old mAs × (old screen speed/new screen speed), 5 = 20 × (200/800).

77. **a.** Short-scale contrast refers to a film which has few densities with large differences between them, very black to very white.

78. **c.** A wire mesh test is used to properly ensure film/screen contact.

79. **d.** The crossover rollers are found out of the solutions and help facilitate the movement of film from one solution to the next.

80. **a.** When kilovoltage technique is increased, contrast differences are decreased across the entire image.

81. **c.** Grid ratio is calculated by dividing the height of the lead strips present to the interspace length between them.

82. **b.** When a great amount of scatter strikes a film, it generates areas of decreased density as well as increased fog across the areas that are exposed.

83. **a.** The blue dye manufactured into the base of the film during construction will exhibit density on an unexposed, processed film.

84. **d.** Low kVp techniques produce image contrast that is short scale in nature.

85. **a.** As film/screen speed increases, density will also increase.

86. **c.** Filtration assists in decreasing the dosage by filtering out all useless soft, low energy photons from reaching the patient.

87. d. New mAs = old mAs × (new grid conversion factor/old grid conversion factor), 64 mAs = 32 mAs [6(16:1)/3(8:1)].

88. d. Perceptible increases in density are only visualized when mAs has been increased by 30%.

89. b. Utilizing the 15% rule by increasing the kVp 15%, the density on the image will double. In order to maintain density, one half the mAs must be utilized.

90. d. Scattered radiation is produced after interaction of the beam with the patient.

91. c. Focal spot size has the least effect on patient dosage.

92. a. Replenishment rates are controlled by replenishing the solutions for every 14 inches of film traveling through.

93. a. Quantity of photons generated is commonly referred to as mAs.

94. b. An increase in OID will decrease the detail of an image due to magnification. A decrease in OID will increase the detail of an image.

95. a. Rollers serve to transport a film throughout an automatic processor.

96. b. Increasing photon energy, which is done by increasing the kVp, will decrease the contrast of an image.

97. a. Increasing the mAs from 10 to 20 would increase the density on the image.

98. a. Cold developer solution will react much slower on a film thus resulting in the film appearing under exposed.

99. b. In order to double the density, kVp must be increased 15%; to halve the density, it should be decreased 15%.

100. a. Angling the x-ray tube to the part will elongate the image.

Radiographic Procedures

101. d. General conventional tomography utilizes the motion of the x-ray tube and the IR while the part being imaged remains still in order to visualize a specific plane of the part.

102. d. IOML and the MSP should be parallel with the plane of the IR; IPL should be perpendicular.

103. b. The lower margin of the IR should be at the level of the iliac crest for the lower rib series.

104. c. The dens and the lateral masses should be visualized within the open mouth.

105. a. The lateral wall of the right ascending colon should be visualized in the left lateral decubitis projection as the contrast layers down.

106. b. The lower leg is internally rotated for an AP hip projection in order to properly visualize the femoral neck free of superimposition.

107. c. With the wrist in ulnar flexion, the scaphoid is projected free of superimposition.

108. b. The sternal (jugular) notch is located at the level of T-2/T-3.

109. d. Sinus studies should be performed in the erect position to properly visualize air/fluid levels within.

110. b. The central ray should be directed two inches superior to the EAM for a lateral skull image.

111. a. The liver takes up the vast majority of the right upper quadrant of the abdomen.

112. d. The Trendelenberg position is described by the feet elevated above the head.

113. a. The eye of the "Scottie dog" is the pedicle.

114. b. The degree of obliquity for projections of the SI joints is 25 degrees.

115. c. A hangman's fracture refers to a complete transverse fracture of the dens.

116. c. The AP 15 degree medial oblique of the ankle properly demonstrates the ankle mortise.

117. d. Heartbeat, peristalsis, and blood flow are considered involuntary motions.

118. d. The AP oblique of the wrist demonstrates the pisiform free of superimposition from the other carpal bones.

119. b. The liver is the site of bile production.

120. c. The pancreas is the site of insulin production and regulation.

121. b. Peristalsis is slowest at the more distal part of the gastrointestinal tract, hence it is slowest at the colon.

122. b. The pyloric valve is the opening between the pylorus of the stomach and the duodenum of the small intestine.

123. a. A right lateral decubitus view will demonstrate a pneumothorax on the left side, as the free air will rise in the thoracic cavity.

124. b. The lower margin of the cassette should be just below the level of the symphysis pubis in order to visualize the entire bladder.

125. d. This position described is the AP Townes Projection for occipital bone.

126. b. The ileocecal valve is the area where the small intestine terminates into the colon; therefore, when contrast media empties through this area during a dedicated small bowel series, the test is completed.

127. b. Adduction is when a part moves toward the midline of the body.

128. c. The RAO position is best utilized for the sternum in order to free it from superimposition of the thoracic spine, as well as utilizing the heart shadow as contrast to better visualize.

129. b. The tibial tuberosity is the raised surface on the anterior surface of the proximal tibia.

130. c. In order to best view the glenoid cavity in profile, the Grashey Projection, which is an AP 45 degree obliquity with the side of interest toward the IR, is to be used.

131. c. When bilateral hip fractures are suspected, the Clements-Nakayama Projection should be utilized in conjunction with an AP projection for proper diagnosis.

132. c. Bilateral Pearson Projections for AC joints are done in the erect position with and without the patient holding weights.

133. d. There are 8 carpal bones in one wrist; therefore, there are 16 carpal bones in both wrists.

134. c. Facial bone studies should be done in the erect position due to the likelihood of blood pooling in the sinuses, which would help diagnose a small fracture.

135. d. The thoracic spine presents a kyphotic curve.

136. b. For the right sacroiliac joint, the patient can be positioned either 25 degrees LPO or RAO for visualization.

137. d. In the cervical vertebra, the zygapophyseal joints are visualized in the lateral projection.

138. c. The Rheese Projection of the orbits is utilized to view blowout fractures.

139. a. Carpal interspaces of the wrist are best visualized using the AP projection.

140. b. The cassette should be placed in a vertical device with its upper margin at iliac crest and angled 15 degrees to be parallel with the femoral neck.

141. a. The cervical vertebra contains a foramen in its transverse processes for travel of the jugular veins.

142. c. This describes the orbitomeatal line.

143. c. The AP projection with the humerus in external rotation will view the greater tuberosity in profile on the lateral margin.

144. b. The talus articulates with the calcaneous distally.

145. c. Breathing should be suspended at the end of full exhalation in order to visualize the lower ribs utilizing the contrast of the abdomen.

146. b. The navicular is the most frequently fractured carpal bone from falls.

147. b. The central ray should be angled 5 degrees cephalic in order to elevate the medial femoral condyle out of the joint space on a lateral knee image.

148. c. The sphenoid sinuses are projected through the mouth in the modified Waters Projection.

149. b. The central ray should be directed 15 degrees cephalic for an AP projection of the sacrum.

150. c. There should be ten posterior ribs above the diaphragm on a properly exposed chest radiograph.

151. c. The cranium is composed of eight bones.

152. d. The cauda equina, or termination of the spinal cord, is at the level of L-1/L-2.

153. a. The mass of the liver keeps the organs on the right side of the abdomen depressed when compared to the left.

154. c. The humerus is the largest bone in the upper extremity.

155. a. The pituitary gland lies within the head cradled by the sella turcica.

156. a. An asthenic body is described by a frail build with a shallow and narrow thorax.

157. c. The degree of rotation for a scapular "Y" view is generally 60 degrees.

158. c. The CR of a PA hand projection should be directed to the third metacarpophalangeal joint.

159. d. In order to reduce the natural OID of the wrist, the fingers should be flexed in order to bring the carpals closer to the plane of the IR.

160. b. In the case that a patient cannot extend the elbow, two AP projections with partial flexion for the distal humerus and proximal forearm should be performed.

Radiation Protection

161. b. Size of cell does not influence radiation effect.

162. a. One dose delivery causes the most profound effect.

163. a. High kVp can allow lower mAs dosages.

164. c. The size of field prevalently affects the radiation amounts received during diagnostic studies.

165. c. Compton scatter interactions are most common in tissue at 100 kVp.

166. c. 2.5 mm aluminum equivalent is the general purpose radiography units' minimal filtration thickness.

167. c. Twelve inches is the minimum distance.

168. b. Coulomb per kilogram (C/Kg) is used to measure ionization in air.

169. a. The later the better, as cells would be more mature.

170. d. Filtration removes weaker (longer wavelength) photons.

171. c. Ionization is the effect, biologically speaking.

172. d. All of these matter to the degree to which recovery is possible.

173. a. Alpha has high LET, but it lacks penetrability.

174. d. LET stands for linear energy transfer.

175. c. There are two main classes: somatic and germ cells.

176. b. Mitosis division is related to somatic cells.

177. a. The DNA molecule is struck in the direct hit theory.

178. c. The double helix is a DNA molecule.

179. d. All of the damages can occur.

180. c. DNA stands for deoxyribonucleic acid.

181. a. Oxygen enhances radiosensitivity.

182. d. Irradiation of bone marrow affects white blood cells.

183. a. RBE stands for relative biologic effectiveness.

184. b. Inverse square law is about distance.

185. b. Radiolysis is water/hydrogen/oxygen.

186. a. Interphase death is apoptosis.

187. a. Sublethal death does not kill the cell.

188. c. Patient skin exposure is improved with filtration.

189. b. With most fluoroscopy units, the tube is in the table, shooting up.

190. d. The dormant stage is not among the major steps.

191. b. Cancer is a late effect.

192. c. Decreasing the SID and increasing shielding follow the basic principals.

193. c. Primary barriers must be at least 8 feet high.

194. a. Radiation monitors must be worn at the collar, outside of the lead to monitor the thyroid dose.

195. b. The allowable dose limit is 5 rems.

196. b. The person will receive 9.5 R/min according to the inverse square law.

197. c. Roentgen is the measurement of radiation in the air.

198. b. It stands for as low as reasonably achievable.

199. d. Gonad shielding will absorb 95%.

200. d. It is never acceptable for a technologist to hold a patient.

APPENDIX

State Licensing Contacts

ARIZONA
MRTBE
4814 S. 40th St.
Phoenix, AZ 85040

602-255-4845
www.azrra.gov

ARKANSAS
Arkansas Dept. of Health
Radiologic Technology Licensure Program
Freeway Medical Bldg.
5800 W. 10th St., Ste. 100
Little Rock, AR 72204

501-661-2166
www.healthyarkansas.com/rtl/index.html

CALIFORNIA
California Dept. of Public Health
Radiologic Health Branch Certification
PO Box 997414, MS #7610
Sacramento, CA 95899-7414

916-327-5106
www.cdph.ca.gov

COLORADO
Colorado Dept. of Public Health and Environment
HMWMD—Radiation Control Program
X-Ray Certification Unit
4300 Cherry Creek Dr. S., #B2
Denver, CO 80246-1530

303-692-3448
www.cdphe.state.co.us

CONNECTICUT
Connecticut Dept. of Public Health
Radiographer Licensure
410 Capitol Ave., MS #12APP
PO Box 340308
Hartford, CT 06134

860-509-7603
www.ct.gov/dph

DELAWARE
Delaware Division of Public Health
Office of Radiation Control
417 Federal St.
Dover, DE 19901

302-744-4546
www.deph.org

FLORIDA

Florida Dept. of Health
Certification Office for EMT/Paramedic/RAD Tech/RA/
Med Phys.
4052 Bald Cypress Way, Bin C85
Tallahassee, FL 32399-3285

850-245-4910
www.doh.state.fl.us/mqa/Rad-Tech

HAWAII

Hawaii Radiologic Technology Board
591 Ala Moana Blvd., Rm. #133
Honolulu, HI 96813-4921

808-586-4700
www.hawaii.gov/health/permits/ehsd/index-ehsd.html

ILLINOIS

IEMA
1035 Outer Park Dr.
Springfield, IL 62704

217-785-9913
www.iema.illinois.gov/radiation/accredit.asp

INDIANA

Indiana State Dept. of Health
Medical Radiology Services 5F
2 North Meridian St.
Indianapolis, IN 46204-3003

317-233-7565
www.in.gov/isdh/23279.htm

IOWA

Iowa Dept. of Public Health
Bureau of Radiological Health
Lucas State Office Building, 5th Floor
321 East 12th St.
Des Moines, IA 50319

515-281-0415
www.idph.state.ia.us

KANSAS

Kansas State Board of Healing Arts
235 S. Topeka Blvd.
Topeka, KS 66603

888-886-7205
www.ksbha.org

KENTUCKY

Radiation Health Branch
275 East Main St.
HS 1 C-A
Frankfort, KY 40621

502-564-3700
www.chfs.ky.gov/dph/radiation

LOUISIANA

Louisiana State Radiologic Technology Board of Examiners
3108 Cleary Ave., Ste. 207
Metairie, LA 70002

504-838-5231
www.lsrtbe.org

MAINE

Radiologic Technology Board of Examiners
State House Station #35
Augusta, ME 04333

207-624-8626 or 207-624-8603
www.maine.gov/professionallicensing

MARYLAND

Maryland Board of Physicians
4201 Patterson Ave.
PO Box 2571
Baltimore, MD 21215-0002

410-764-4777
www.mbp.state.md.us/pages/info.html

MASSACHUSETTS

MA Dept. of Public Health
Radiation Control Program
Schrafft Center, Ste 1M2A
529 Main St.
Charlestown, MA 02129

617-242-3035 x2005
www.mass.gov/dph/rcp

MINNESOTA

Minnesota Dept. of Health
Indoor Environments and Radiation Section, X-Ray Unit
PO Box 64497
St. Paul, MN 55164-0497

651-201-4545
www.health.state.mn.us/xray

MISSISSIPPI

Mississippi State Dept. of Health
Professional Licensure
PO Box 1700
Jackson, MS 39215

601-364-7360
www.healthyms.com

MONTANA

Board of Radiologic Technologists
301 S. Park, 4th Floor
PO Box 200513
Helena, MT 59620

406-841-2385
www.radiology.mt.gov

NEBRASKA

DHHS
Licensure Unit
PO Box 94986
Lincoln, NE 68509-4986

402-471-2118
www.dhhs.ne.gov

NEW JERSEY

New Jersey Dept. of Environmental Protection
Bureau of Radiological Health
PO Box 415
Trenton, NJ 08625

609-984-5890
www.nj.gov/dep/rpp

NEW MEXICO

NM Radiologic Technology Program
Radiation Control Bureau
PO Box 26110
Santa Fe, NM 87502-6110

505-476-3264
www.nmenv.state.nm.us/nmrcb/radtech.html

NEW YORK

BERP—Radiologic Technology
New York Dept. of Health
547 River St., Rm. 530
Troy, NY 12180

518-402-7580
www.nyhealth.gov/radiation

OHIO

Ohio Dept. of Health
Radiologic Technology Section
246 North High St.
Columbus, OH 42315

614-752-4319
www.odh.ohio.gov

OREGON

Oregon Board of Radiologic Technology
800 NE Oregon St., Ste. 1160A
Portland, OR 97232-2162

971-673-0215
www.oregon.gov/RadTech

PENNSYLVANIA

State Board of Medicine
State Board of Osteopathic Medicine
PO Box 2649
Harrisburg, PA 17105

Medicine: 717-783-1400
Osteopathic: 717-783-4858
Medicine: www.dos.state.pa.us/med
Osteopathic: www.dos.state.pa.us/ost

RHODE ISLAND

Radiologic Board
Rhode Island Dept. of Health
3 Capitol Hill
Providence, RI 02908

401-222-2839
www.health.ri.gov

SOUTH CAROLINA

South Carolina Radiation Quality Standards Association
PO Box 7515
Columbia, SC 29202

803-771-6141
www.scrqsa.org

TENNESSEE

Tennessee Board of Medical Examiners
227 French Landing, Ste. 300
Heritage Place
MetroCenter
Nashville, TN 37243

615-532-4384
health.state.tn.us/boards/xrayop

Tennessee Examination Processing Center
PO Box 41776
Nashville, TN 37204

615-383-9499
www.limitedscope.com

TEXAS

Medical Radiologic Technologist Program
Dept. of State Health Services
MS 1982
PO Box 149347
Austin, TX 78714-9347

512-834-6617
www.dshs.state.tx.us/mrt

UTAH

Division of Occupational and Professional Licensing
160 E. 300 S.
PO Box 146741
Salt Lake City, UT 84114-6741

801-530-6628
www.dopl.utah.gov/licensing/radiology.html

Utah Exam Processing Center
PSI Exams LLC
2950 N. Hollywood Way, Ste. 200
Burbank, CA 91505

800-733-9267
candidate.psiexams.com/index.jsp

VERMONT

Board of Radiology
Office of Professional Regulation
National Life Bldg. North, Floor 2
Montpelier, VT 05620-3402

802-828-2373
www.vtprofessionals.org

VIRGINIA

Commonwealth of Virginia
Dept. of Health Professions
9960 Mayland Dr., #300 Perimeter Center
Henrico, VA 23233-1463

804-367-4501
www.dhp.virginia.gov

WASHINGTON

Dept. of Health
Naturopathy Program
PO Box 47852
Olympia, WA 98504-7852

360-236-4700
www.doh.wa.gov/hsqa/professions/Radiological_
Technologist/default.htm

WEST VIRGINIA

WV Medical Imaging and Radiation Therapy Technology
Board of Examiners
PO Box 638
1715 Flat Top Rd.
Cool Ridge, WV 25825-0638

304-787-4398
www.wvrtboard.org

WYOMING

Wyoming Board of Radiologic Technologist Examiners
1800 Carey Ave., 4th Floor
Cheyenne, WY 82002

307-777-3507
plboards.state.wy.us/radiology/index.asp

Professional Societies

American Academy of Oral and Maxillofacial Radiology (AAOMR)

American Association of Physicists in Medicine (AAPM)

American Association for Women Radiologists (AAWR)

American Board of Nuclear Medicine (ABNM)

American Board of Radiology (ABR)

American College of Forensic Examiners Institute (ACFEI)

American College of Medical Physics (ACMP)

American College of Nuclear Physicians (ACNP)

American College of Radiation Oncology (ACRO)

American College of Radiology (ACR)

American College of Healthcare Executives (ACHE)

American Healthcare Radiology Administrators (AHRA)

American Institute of Ultrasound in Medicine (AIUM)

American Medical Association (AMA)

American Nuclear Society (ANS)

American Osteopathic College of Radiology (AOCR)

American Radium Society (ARS)

American Registry of Diagnostic Medical Sonographers (ARDMS)

American Registry of Radiologic Technologists (ARRT)

American Roentgen Ray Society (ARRS)

American Society for Gastrointestinal Endoscopy (ASGE)

American Society for Therapeutic Radiology and Oncology (ASTRO)

American Society of Echocardiography (ASE)

American Society of Emergency Radiology (ASER)

American Society of Head and Neck Radiology (ASHNR)

American Society of Interventional and Therapeutic Neuroradiology (ASITN)

American Society of Neuroradiology (ASNR)

American Society of Nuclear Cardiology (ASNC)

American Society of Radiologic Technologists (ASRT)

American Society of Spine Radiology (ASSR)

American Telemedicine Association (ATA)

Association of Residents in Radiation Oncology (ARRO)

Association of Telemedicine Service Providers (ATSP)

Association of University Radiologists (AUR)

Association of Vascular and Interventional Radiographers (AVIR)

Canadian Association of Radiologists (CAR)

Canadian Organization of Medical Physicists (COMP)

Clinical Magnetic Resonance Society (CMRS)

Computer Assisted Radiology and Surgery (CARS)

International Association of Dentomaxillofacial Radiology (IADMFR)

International Consortium for Medical Imaging Technology (ICMIT)

International Society for Magnetic Resonance in Medicine (ISMRM)

International Society of Radiographers and Radiological Technologists (RSRRT)

International Society of Radiology (ISR)

International Society of Ultrasound in Obstetrics and Gynecology (ISUOG)

Musculoskeletal Ultrasound Society

New York State Society of Radiologic Sciences (NYSSRS)

North American Society for Cardiac Imaging (NASCI)

Nuclear Medicine Technology Certification Board (NMTCB)

Radiological Society of North America (RSNA)

Radiology Business Management Association (RBMA)

Radiation Research Society (RRS)

Radiation Therapy Oncology Group (RTOG)

Society of Breast Imaging (SBI)

Society for the Advancement of Women's Imaging (SAWI)

Society for Cardiac Angiography and Interventions (SCAI)

Society for Cardiovascular Magnetic Resonance

Society for Computer Application in Radiology (SCAR)

Society for Pediatric Radiology (SPR)

Society of Cardiovascular and Interventional Radiology (SCVIR)

Society of Cardiovascular Magnetic Resonance (SCMR)

Society of Computed Body Tomography and Magnetic Resource (SCBT/MR)

Society of Diagnostic Medical Sonographers (SDMS)

Society of Gastrointestinal Radiologists (SGR)

Society of Nuclear Medicine (SNM)

Society of Radiologists in Ultrasound (SRU)

Society of Radiology Oncology Administrators (SROA)

Society of Skeletal Radiology (SSR)

Society of Thoracic Radiology (STR)

Society of Uroradiology (SUR)

GLOSSARY

abduction Movement of body part away from the midline..

absorption Influenced by the thickness of the body part and the type (atomic number); the higher of either, the more absorption and the higher the production of tissue of scatter radiation.

acanthiomeatal line (aml) Runs from the acanthion to the eam.

acanthion The tip of the nose.

adduction Movement of the body part toward the midline.

advance directive Document to be filled out by patient to provide directives regarding medical care before he or she becomes incapacitated.

airborne precautions Respiratory protection required for individuals entering patient's room.

airborne transmission Droplets and dust.

air-gap technique Purposely increasing oid so some scattered radiation scatters away from the receptor to provide higher contrast.

ALARA Stands for *as low as reasonably achievable*; tube current is measured in milliamperage.

allergic shock (anaphylaxis) Allergic reaction to foreign proteins following injection of an iodinated contrast agent.

anaphylactic reactions Flushing, hives, nausea.

anatomical position Standing, erect with palms facing outward and all anterior surfaces of the body facing forward.

anode heel effect More photons present on the cathode side of the x-ray tube than on the anode side.

anode stem Made of molybdenum.

aperture diaphragm Beam-restricting device that is a simple, flat, lead diaphragm with an opening cut into its center.

aqueous iodine compound Water-soluble sterile contrast agent.

assault The threat of injurious touching.

attenuation The reduction of intensification as a result of absorption and scattering.

auricle Exterior portion of the ear.

automatic exposure controls   Type of device that terminates an exposure when the system has determined that a certain measured exposure reaches the image receptor; it allows for manual manipulation of mA and kVp.

axial   Angling of the central ray greater than 10 degrees.

battery   Unlawful touching of a person without his or her consent.

beam limiting devices   Devices located between the anode and cathode and the patient that restrict the size of the field being irradiated; cones and collimators are prime examples; other than limit the field, which is a primary radiation protection device, these devices improve visibility of detail by eliminating photons for prevention of potential scatter radiation.

blood pressure   Normal adult blood pressure is 120/80 mm Hg.

bradycardia   Medical condition where a patient presents with fewer than 60 heartbeats per minute.

cardiac arrest   Cessation of heart function.

cardiogenic shock   Secondary to cardiac failure or other interference with heart function.

cardiovascular reactions   Hypotension, tachycardia, cardiac arrest.

cathode/filament   Negative terminal of tube; supplies the electrons when heated.

characteristic curve   Also referred to as H and D curve and sensitometric curve, this measures the optical density on a film and can be utilized to measure speed, latitude, and contrast in film.

chest tube   In place to remove fluid or air from the pleural space.

coherent (classical) scatter   Scattering of x-rays with no loss of energy; this is also referred to as Rayleigh or Thompson scattering.

collimator   Most complex beam-restricting device that allows for unlimited restriction sizes and also provides a light source.

common vehicle transmission   Primarily, transmission of infectious organisms by contaminated items such as food, water, medications, devices, and equipment.

compensating filter   Device utilized to compensate when there are varying thicknesses in the part being imaged; this will produce a more uniform density across the image.

Compton effect   The scattering of x-rays resulting in ionization and loss of energy.

cone   Beam-restricting device that is an aperture diaphragm with an extension that flares outward.

contact precautions   Masks, gloves, and gowns are indicated for individuals coming in contact with a potentially infectious patient.

crash cart   Used in cardiac arrest, this cart contains medications, airways, sphygmomanometers, stethoscopes, defibrillators, and cardiac monitors.

crossed-hatch grid   A type of grid in which two parallel grids are "interlaced" together and the grid strips of each grid are positioned perpendicular to one another.

current   The amount of electricity per second through the tube.

cylinder   Beam-restricting device consisting of an aperture diaphragm with an extension that continues straight.

developer   Alkali solution that reduces the silver ions present at the sensitivity speck into black metallic silver.

diastolic pressure   Measurement of the heart at rest.

DICOM   Acronym for digital imaging and communications in medicine.

direct contact transmission   Infected person touches a susceptible host.

diversity   Factors that differentiate humans from one another; factors include age, gender, race or ethnicity, sexual preference, marital status, religious beliefs, etc.

do not resuscitate order (DNR)   Patient chooses to waive a code response.

dorsal decubitus   Patient supine with central ray passing horizontally.

dorsiflexion   Flexion between the lower leg and the foot, ultimately decreasing the angle between the two body parts to less than 90 degrees.

droplet precautions   Masks are required for persons coming in close contact with the patient.

droplet transmission   Primarily transmission by coughs, sneezes, or other methods of spraying onto a nearby host.

durable power of attorney   Patient allows for another person to make decisions regarding medical care if he or she can no longer communicate.

effective atomic number   When the effective atomic number of adjacent tissues is very high, subject contrast is very high.

elongation   Object appears longer than actual on a radiograph.

eversion   Turning the foot outward at the ankle joint.

extension   Increasing the angle of the joint by extending or straightening the joint; the term *hyperextension* is extending the joint beyond its normal limitations.

external acoustic meatus (EAM)   The passageway within the ear between the ear flap and the eardrum.

false imprisonment   A detention of a person against his or her will.

filament current   The current needed to heat the filament, measured in amperage.

filter   Material, commonly aluminum, which assists in removing low-energy photons that serve little diagnostic purpose.

fixer   An acidic solution that stops the developing process, removes the unexposed silver bromide crystals from the film, and preserves and hardens the film.

flexion   Decrease in the angle of the joint by the movement of bending; the term *hyperflexion* is equal to *overflexion*.

focal-spot blur   Also known as penumbra, blur that appears on the periphery of the image due to the focal spot.

focusing cup   Reduces the spreading of the space charge.

focused grid   Constructed so that the grid strips are angled to coincide with the divergence of the x-ray beam from its source at a given distance, which should be noted on the grid itself.

foreshortening   Object appears smaller than actual on a radiograph.

Fowlers   Head elevated 45–90 degrees.

glabella   Located in the space between the eyebrows, merger of the two superciliary ridges.

glabelloalveolar line   Passes through the anterior portion of the glabella, nasion, and the acanthion when in a lateral position.

glabellomeatal line   Runs from the glabella to the eam.

glass tube (Pyrex)   Holds the cathode and anode and keeps vacuum intact.

gonion   Angle of mandible.

gray   (Gy) international standard (SI) for the absorbed dose, equal to the rad.

grid   Device made up of thin strips of radiopaque material (usually lead) separated by a radiolucent interspace encased by a thin cover of aluminum that is designed to effectively reduce the amount of scatter reaching the image receptor.

grid frequency   Refers to the number of grid strips per centimeter.

grid ratio   The height of the lead strip divided by the interspace.

half value layer (hvl)   The thickness of absorbing material necessary in order to reduce the intensity of the beam to half of its original intensity.

hand washing   The most effective method to prevent the spread of infection.

HIPAA   Acronym for the Health Insurance Portability and Accountability Act.

hypodermic needle gauge   Unit of measurement that indicates diameter; the larger the gauge, the smaller the diameter of the needle opening.

hypovolemic shock   Follows the loss of a large amount of blood or plasma.

implied consent   Assumption of approval of care if the patient was conscious.

indirect contact transmission   Inanimate object containing pathogenic organisms is placed in contact with a susceptible person.

infection control barriers   Gloves, protective clothing, masks, and eye protection.

informed consent   Patient provides consent after being appropriately informed of benefits, risks, need, and alternatives.

infraorbital margin   Inferior portion of the orbit.

infraorbital meatal line (ioml)   Slightly inferior to the oml; it runs from infraorbital margin to the eam.

inherent filtration   Pertaining to the glass envelope, metal enclosure, or dielectric oil surrounding the tube.

inion   The occipital protuberance.

inner cantus   The most medial, inner portion of the eye (almond).

intensifying screens   Material used in conjunction with a matched type of film that is found on the inside part of the cassette; converts x-ray into light photons in order to assist in lowering the dosage necessary in the creation of a diagnostic image.

interpupillary line   This line connects the centers of the orbits and is at a 90 degree angle to the median sagittal plane.

invasion of privacy   The breaching of patient confidentiality or unnecessary touching or exposing a patient's body.

inverse square law $\frac{I_1}{I_2} = \frac{D_2^2}{D_1^2}$ represents the relationship of the intensity of the exposure to distance.

inversion   Turning the foot inward at the ankle joint.

iodinated ionic contrast agents   Salts of organic iodine compounds, composed of positively and negatively charged ions.

iodinated nonionic contrast agents   Do not ionize into separate positive and negative charges.

iv catheter   Combination unit with a needle inside a flexible plastic catheter.

Kilovoltage control   Controls voltage; the unit is the kilovolt.

kilovoltage peak (kvp)   Technical factor controlled by the technologist; it is the maximum voltage applied to an x-ray tube that in turn determines the energy of electrons traveling from cathode to target anode; it controls quality and is the primary controller of contrast; however, it does affect density.

lateral decubitus   Patient lies on either left or right side with the central ray passing either AP or PA.

law of reciprocity   Density of an image will remain constant as long as the mAs value is constant, regardless of the combination of mA and time.

lead   Used as the barrier for protection of x-radiation.

leakage radiation   Radiation which escapes from the tube housing, usually traveling in a different direction than that of the primary beam; remember that when x-rays are created within the tube, the direction they travel in is of an isotrophic nature (in all directions).

left anterior oblique (lao)   Left anterior side of the body closest to the table.

left posterior oblique (Lpo)   Left posterior side closest to the table.

length of exposure   The duration of the exposure in seconds.

level   Height of the histogram, allows for the drawing out or muting of grays.

libel   Documented defamation of character.

linear energy transfer (LET)   As energy is transferred onto and passes through soft tissue, it will deposit its energy as it traverses through that soft tissue; the rate at which it does this is called linear energy transfer.

low contrast   Also referred to as long scale, small differences in adjacent densities; many shades of gray are present on the image.

magnification   Size distortion; image appears larger than the actual object on a radiograph, but it is correctly proportioned.

material safety data sheets (msds)   Provide direction for handling precautions, safe use of the product, cleanup, and disposal of biohazardous materials.

medical asepsis   Microorganisms have been eliminated as much as possible.

mental point   The most anterior point of the mandible in the midline and the most prominent point on the chin.

mentomeatal line   Runs from the mental point to the eam.

midsagittal plane   Separates the head into two symmetrical halves when viewed anteriorly.

milli   Prefix which stands for one-thousandth; for example, a milligray = 1/1,000 of a gray.

milliamperage control   Controls current; the unit is the milliampere.

milliamperage per second (mas)   Technical factor controlled by the technologist; represents the amount of current provided to the cathode filament over a period of time, which in turn provides the number of electrons sent across to the target anode; it controls quantity and is the primary controller of density.

nasion   Intersection of the frontal and two nasal bones, located just superior to the bridge of the nose.

nasogastric (ng) tubes   Tube inserted through the nose and down the esophagus into the stomach.

negative contrast agent   Most commonly used agent is air.

negligence   Neglect or omission of reasonable care and caution.

neurogenic shock   Causes blood to pool in peripheral vessels.

nonstochastic effect   The best example of this effect is if everyone in a group were to be irradiated, everyone would receive an effect, such as cataracts or blood changes (lymphocyte reduction).

nonverbal communication   Communicating with eye contact, facial expressions, or body motions.

OID   Surface of object being imaged-to-image receptor distance.

optical density   Overall degree of blackening on a radiograph.

orbitalmeatal line (oml)   The original "baseline" that runs from the nasion through the outer canthus of the eye to the center of the external auditory meatus.

Outer canthus   the most lateral, outer portion of the eye (almond).

oxygen administration   Usual oxygen flow rate is 3 to 5 L per minute.

PACS   Picture archiving and communication systems.

parallel grid   Type of grid in which all grid strips are parallel to each other.

patient bill of rights   Establishes the rights for patients.

patient education   Providing the patient with appropriate and adequate information regarding the procedure to be performed.

patient history   Provides information about the patient's condition, allergies, and any previous examination they may have had.

phosphor   A material that will emit light photons when stimulated by radiation.

photoelectric absorption   The absorption or energy transfer of x-rays by the process of ionization.

positive beam limitation   Device which automatically adjusts the collimation to correspond with the size of the film loaded.

positive contrast agent   Iodine or barium.

Potter-Bucky diaphragm   Device which assists in the motion of a grid to help blur out radiopaque grid lines; it can be either reciprocating or oscillating.

primary beam   The beam as defined as it travels out of the tube and housing, up until the point when it interacts with the living tissue.

primary radiation   Any photon emerging from the tube that has not yet interacted with tissue.

pronation   Turning of the hand so the palm is down.

prone   Lying face down.

pulse   Normal adult pulse is 60 beats per minute.

RAD   Stands for radiation absorbed dose.

radiographic contrast   Differences in optical densities on an image, shades of gray, affected by the subject and image receptor.

RBE   Stands for relative biologic effect; the comparative measurement of how different types of radiation compare with x- and gamma rays.

recorded detail   Degree of sharpness of the object recorded on the image.

REM   Stands for roentgen equivalent man.

remnant photon   All of the x-ray photons that reach their destination (the film) after passing through the object being radiographed.

remnant radiation   All of the x-ray photons that reach their destination (the film) after passing through the object being radiographed; the photons that have contributed to the image.

res ipsa loquitur   Obvious cause of negligence.

respiration   Normal adult rate is 12 to 16 breaths per minute.

respiratory arrest   Cessation of breathing.

respondeat superior   The employer is responsible for his or her employees' actions.

right anterior oblique (RAO)   Right anterior side of the body closest to the film.

right lateral recumbent position   Erect or recumbent position (lateral), 90 degrees from true AP or PA.

right posterior oblique (RPO)   Right posterior side of patient closet to the table.

roentgen (R)   Named after the discoverer of x-radiation, the roentgen unit is the measurement of exposure of 2.58×10^4 coulombs per kilogram of dry air.

rotating anode (disc)   Attracts the electrons and produces x-rays and positively charged photons.

rotation   Circular motion around a specific axis.

rotor   A copper shaft inside the glass tube that spins when induction occurs through interaction of the stator and the external stator.

scatter radiation   The process in which the x-ray photons undergo a change of direction after interacting with the atoms of an object.

secondary beam   As soon as the primary beam interacts with the living tissue, it is identified as a secondary beam.

shielding   Used to protect individuals from radiation.

sievert (Sv)   International standard for the rem (dose equivalent).

space charge   The electron cloud near the filament.

space charge effect   The repelling of electrons near the filament.

stator   Located outside the glass tube and part of the induction motor, it is a series of electromagnets equally spaced around the neck of the tube.

septic shock   Occurs when toxins produced during massive infection cause a dramatic drop in blood pressure.

shape distortion   Unequal magnification of different parts of the same object due to differing distances being present along the object.

shock   Failure of circulation in which blood pressure is inadequate to oxygenate tissues and remove by-products of metabolism.

SID   Source-to-image receptor distance.

slander   Verbalized defamation of character.

SOD   Source-to-surface distance of the object being imaged.

Spatial distortion   Multiple objects are generally positioned at differing OIDs, causing superimposition or nonvisualization of an object.

spatial resolution   Level of detail.

speed   A screen's or film's ability to convert an amount of radiation to generate an image.

sphygmomanometer   Device used to measure blood pressure.

standard precautions   First tier of transmission-based isolation precautions; use of barriers to prevent contact with blood, all body fluids, non-intact skin, and mucous membranes when there is a chance infection will be transmitted.

stochastic effects   Random in nature; this means that of a group of individuals who are irradiated, some will receive an effect, such as cancer or a genetic abnormal effect, and some will not.

suction unit   Used to maintain patient's airway.

supination   Turning of the hand so the palm is facing upward.

supine   Lying down flat on your back.

surgical asepsis   Complete removal of all organisms from equipment and environment.

systolic pressure   Measurement of the pumping action of the heart.

tachycardia   Medical condition where a patient presents with more than 100 heartbeats per minute.

tangential   To skim the surface of a part.

target of anode   The area of the anode face that the electrons strike during an exposure.

temperature   Normal adult oral temperature is 98.6°F.

thermionic emission   The boiling off of electrons by heating the filament.

tilt   Movement in which the sagittal plane is not parallel to the long axis of the body part.

timer   The manual timer controls the length of the exposure measured in seconds (abbreviated s).

tissue mass density   Different sections of the body mass have equal thicknesses yet different mass densities.

torts   Personal injury law.

trauma   Serious and potentially life-threatening injuries.

Trendelenburg   Patient is supine with head lower than feet.

tube current   The current needed to produce the electrons, which are needed to produce x-ray photons; tube current is measured in milliamperage.

vector-borne transmission   An animal contains and transmits an infectious organism to humans.

ventilators   Mechanical respirators attached to tracheostomies.

ventral decubitus   Patient is prone with central ray passing horizontally.

verbal communication   Communicating by speaking clearly, using precise language understood by the patient.

voltage   The "push" which causes electricity to flow; it is also called the electromotive force (EMF).

ADDITIONAL ONLINE PRACTICE ▶

Whether you need help building basic skills or preparing for an exam, visit the LearningExpress Practice Center! Using the code below, you'll be able to access additional online practice. This online practice will also provide you with:

- **Immediate scoring**
- **Detailed answer explanations**
- **Personalized recommendations for further practice and study**

Log in to the LearningExpress Practice Center by using this URL: **www.learnatest.com/practice**

This is your access code: **7311**

Follow the steps online to redeem your access code. After you've used your access code to register with the site, you will be prompted to create a username and password. For easy reference, record them here:

Username: _____   Password:_____

With your username and password, you can log in and answer these practice questions as many times as you like. If you have any questions or problems, please contact LearningExpress customer service at 1-800-295-9556 ext. 2, or e-mail us at **customerservice@learningexpressllc.com**.

NOTES

NOTES

NOTES

NOTES

NOTES